# Rapid Manufacturing

# Rapid Manufacturing

## An Industrial Revolution for the Digital Age

Editors
**N. Hopkinson, R.J.M. Hague and P.M. Dickens**
Loughborough University, UK

John Wiley & Sons, Ltd

Telephone (+44) 1243 779777

Email (for orders and customer service enquiries): cs-books@wiley.co.uk
Visit our Home Page on www.wiley.com

Reprinted August 2006



***Other Wiley Editorial Offices***

John Wiley & Sons Inc., 111 River Street, Hoboken, NJ 07030, USA

Jossey-Bass, 989 Market Street, San Francisco, CA 94103-1741, USA

Wiley-VCH Verlag GmbH, Boschstr. 12, D-69469 Weinheim, Germany

John Wiley & Sons Australia Ltd, 42 McDougall Street, Milton, Queensland 4064, Australia

John Wiley & Sons (Asia) Pte Ltd, 2 Clementi Loop #02-01, Jin Xing Distripark, Singapore 129809

John Wiley & Sons Canada Ltd, 22 Worcester Road, Etobicoke, Ontario, Canada M9W 1L1

Wiley also publishes its books in a variety of electronic formats. Some content that appears in print may not be available in electronic books.

***Library of Congress Cataloging-in-Publication Data***

Rapid manufacturing: an industrial revolution for the digital age/
edited by N. Hopkinson, R.J.M. Hague, and P.M. Dickens.
p. cm.
Includes bibliographical references and index.
ISBN-13 978-0-470-01613-8 (cloth : alk. paper)
ISBN-10 0-470-01613-2 (cloth : alk. paper)
1. Rapid protoyping. 2. Rapid tolling. I. Hopkinson, N. II.
Hague, R. J. M. III. Dickens, P. M.
TS171.4.R34 2005
620′.0042′0285–dc22 2005018264

***British Library Cataloguing in Publication Data***
A catalogue record for this book is available from the British Library

ISBN-13 978-0-470-01613-8 (H/B)
ISBN-10 0-470-01613-2 (H/B)

Typeset in 11/13pt Palatino by Thomson Press (India) Limited, New Delhi, India.
Printed and bound in Great Britain by Antony Rowe Ltd, Chippenham, Wiltshire
This book is printed on acid-free paper responsibly manufactured from sustainable forestry in which at least two trees are planted for each one used for paper production.

# Contents

# List of Contributors

**Dave Bourell** Mechanical Engineering, The University of Texas at Austin, 1 University Station Stop C2200, Austin, TX 78712, USA, dbourell@mail.utexas.edu

**Ian Campbell** Department of Design and Technology, Loughborough University, Loughborough, UK, r.i.campbell@lboro.ac.uk

**Phill Dickens** Rapid Manufacturing Research Group, Loughborough University, Loughborough, UK, P.Dickens@lboro.ac.uk

**Poonjolai Erasenthiren** Rapid Manufacturing Research Group, Loughborough University, Loughborough, UK, p.erasenthiren@lboro.ac.uk

**Brad Fox** Center for Rapid Manufacturing, 11195 Kenora Way, Lakeville, MN 55044, USA, rapidinsert@yahoo.com

**Tim Gornet** Rapid Prototyping Center, University of Louisville, Louisville, KY 40292, USA, Tim.gornet@louisville.edu

**Richard Hague** Rapid Manufacturing Research Group, Loughborough University, Loughborough, UK, R.hague@lboro.ac.uk

**Russ Harris** Rapid Manufacturing Research Group, Loughborough University, Loughborough, UK, r.a.harris@lboro.ac.uk

**Neil Hopkinson** Rapid Manufacturing Research Group, Loughborough University, Loughborough, UK, n.hopkinson@lboro.ac.uk

**Rik Knoppers** TNO Science and Industry, Postbus 6235, 5600 HE Eindhoven, The Netherlands, g.knoppers@ind.tno.nl

**Janne Kytannen** Freedom of Creation, Hobbemakade 85 hs, 1071 XP Amsterdam, The Netherlands, janne@freedomofcreation.com

**Fred McBagonluri** Siemens, 10 Constitution Avenue, Piscataway, NJ 08844, USA, fmcbagonluri@siemens.com

**Martin Masters** Siemens, 10 Constitution Avenue, Piscataway, NJ 08844, USA, mmasters@siemens.com

**Monica Savalani** Rapid Manufacturing Research Group, Loughborough University, Loughborough, UK, m.m.savalani@lboro.ac.uk

**Rupert Soar** Rapid Manufacturing Research Group, Loughborough University, Loughborough, UK, r.c.soar@lboro.ac.uk

**Roger Spielman** Technology Development Manager, Solid Concepts, 28231 Avenue Crocker Bldg 10, Valencia, CA 91355, USA, rspielman@solidconcepts.com

**Graham Tromans** Rapid Manufacturing Research Group, Loughborough University, Loughborough, UK, G.P.Tromans@lboro.ac.uk

**Chris Tuck** Rapid Manufacturing Research Group, Loughborough University, Loughborough, UK, c.j.tuck@lboro.ac.uk

**Therese Velde** Siemens, 10 Constitution Avenue, Piscataway, NJ 08844, USA, tvelde@siemens.com

**John Wooten** 625F Ivywood Lane, Simi Valley, CA 93065, USA, jrwooten@sbcglobal.net

# Editors

Neil Hopkinson is a lecturer in the Wolfson School of Mechanical and Manufacturing Engineering at Loughborough University, UK. Having obtained his PhD in Rapid Tooling in 1999, Neil began to look into the economic viability of Rapid Manufacturing. Inspired by the findings of this research Neil began to investigate low-cost, high-speed Rapid Manufacturing processes while also focusing his research on material properties in powder-based layer manufacturing processes. To date Neil has secured over £1 million of research funding and published over 40 journal/conference papers; he was also an invited visiting lecturer at the University of Queensland in Australia.

Phill Dickens is Professor of Manufacturing Technology and Director of the Innovative Manufacturing and Construction Research Centre at Loughborough University, UK. He is also Associate Dean of Research for the Engineering Faculty. Phill started work in the area of Rapid Prototyping in 1990, working on processes such as 3D Welding and using copper coated SL models as electrodes for EDM. The research work has changed emphasis since then from Rapid Prototyping to Rapid Tooling and is now concentrating on Rapid Manufacturing.

Richard Hague is a Senior Lecturer and Head of the Rapid Manufacturing Research Group at Loughborough University, UK. He has been involved with Rapid Prototyping and Rapid Manufacturing (RM) research since 1993, and is now Principal Investigator on several large EPSRC, DTI and EU funded research projects. He was also instrumental in setting up and managing the successful Rapid Manufacturing Consortium that now operates from Loughborough. Richard has many academic publications in the area of Rapid Manufacturing and is referee to several international academic journals and conferences. He also holds a patent that was gained as part of his PhD studies which is licensed to the predominant manufacturer of Rapid Prototyping equipment (3D Systems Inc.).

# Foreword

It is a privilege to write the Foreword for this very important book. Rapid Manufacturing (RM) is the next frontier for researchers, developers and users of a technology that has been used predominately for Rapid Prototyping. The additive, freeform nature of the technology, coupled with improvements in materials, processing speed, accuracy and surface finish, opens up an array of options that before were impossible. In the not too distant future, series production applications of the technology will propel machines sales and the number of parts produced annually to impressive new levels. The numbers for Rapid Prototyping applications will pale in comparison.

RM has a promising future with a powerful list of potential benefits. In fact, a growing number of companies are betting their future on systems for Rapid Manufacturing. The technology makes it feasible to manufacture series production parts economically in quantities of one to several thousand pieces, directly, depending on part size and other factors. Without the constraints imposed by tooling, designers are given the freedom to create new designs that before were impossible or impractical to manufacture. This method of manufacturing presents a dramatically different way of thinking and the future implications are staggering. Personalized products become possible because the cost of producing a unique design is not prohibitively expensive, as it is with tooling or other costly methods of manufacturing such as casting.

True just-in-time manufacturing becomes a reality because companies can produce parts as they are needed, rather than in large batches as they are done today. The relatively simple and automated operation of the machines makes it far easier to decentralize production operations. Placing machines at or near customer sites and sending new designs to them using Internet tools also becomes an interesting option.

Most new designs never see the light of day because they are too risky to manufacture due to the high cost of tooling. When tooling is removed from the equation, it becomes feasible to introduce new products in low quantities to see whether a market demand exists for them. The established method is to spend months, or longer, and hundreds of thousands of euros, or more, to find out whether a new design is a winning product.

Many nations are losing much of their manufacturing base to countries that can produce products much less expensively. RM will help these nations preserve some of their manufacturing because it will become impractical to manufacture low quantities of parts in such far away places. Manufacturing parts locally, and just in time, will become much more attractive than manufacturing them halfway around the world because of the time and cost of shipping. Also, design data that are sensitive will remain inside the company.

The Rapid Manufacturing Research Group at Loughborough University shares this vision. It is the only group of its kind in the world with a staff of academic professionals dedicated exclusively to Rapid Manufacturing. In recent years, the group has expanded to more than 40 researchers who provide a continuous stream of new ideas, new projects and new results. I have had the pleasure of working with and getting to know many of these fine people and I can say without reservation that they are among the top thinkers in the world. They have gained my respect and the respect of countless others.

The Rapid Manufacturing Research Group has secured some of the best people in the industry to contribute their ideas and experience to this book. The collective knowledge and experience of these people has resulted in a publication like no other. Study it, absorb its many ideas and examples, and use the information to form your own opinions about RM's future. Finally, feel fortunate that this group has come together to help advance product development and manufacturing around the world. Congratulations to them for this impressive achievement.

Terry Wohlers
*Wohlers Associates, Inc.*

# 1

# Introduction to Rapid Manufacturing

Neil Hopkinson, Richard Hague and Phill Dickens
*Loughborough University*

## 1.1 Definition of Rapid Manufacturing

The definition of Rapid Manufacturing (RM) can vary greatly depending on whom one talks to. For some people it can simply mean making end-use parts quickly – by any manufacturing method – while for others it involves the use of an additive manufacturing process at some stage in the production chain.

Our definition is very clear and precise. Rapid Manufacturing is defined as 'the use of a computer aided design (CAD)-based automated additive manufacturing process to construct parts that are used directly as finished products or components'. The additive manufactured parts may be post-processed in some way by techniques such as infiltration, bead blasting, painting, plating, etc. The term 'additive' manufacturing is used in preference to 'layer' manufacturing as it is likely that some future RM systems will operate in a multi-axis fashion as opposed to the current layer-wise manufacturing encountered in today's Rapid Prototyping (RP).

Although current RP systems are being successfully used in specialist applications for the production of end-use parts, these RP systems have not been designed for manufacturing and many problems remain to be solved. These include surface finish, accuracy and repeatability, among others. We are currently in a transition stage where RP systems are being used for these specialist, low-volume and customised products but true manufacturing that

*Rapid Manufacturing: An Industrial Revolution for the Digital Age*
Editors N. Hopkinson, R.J.M. Hague and P.M. Dickens 

is of a sufficient speed, cost and quality that can be accepted by the general consumer does not exist at present.

The field of Rapid Manufacturing has grown in recent years and offers such significant potential that it must be considered as a discipline in its own right that is independent from its predecessors of Rapid Prototyping and Rapid Tooling. This new discipline, which eliminates tooling, has profound implications on many aspects of the design, manufacture and sale of new products.

## 1.2 Latitude of Applications

It is difficult to think of a technological discipline that has such a broad range of potential applications as Rapid Manufacturing. What other technology can get an artist, a medical clinician, an engineer and an environmental champion excited in the same way? This almost unparalleled latitude of applications is reflected in the range of materials that may be processed. We are only in the early stages of developing the technologies but are already able to reliably process parts in polymers, metals and ceramics and the potential for functionally graded components adds a degree of freedom for a combination of materials that had not previously existed. Not only is there a great breadth of potential for Rapid Manufacture, but the discipline also brings about issues of significant depth, uncovering new ways of thinking in terms of many aspects ranging from involvement of the customer in the design process through to the ability to realise new engineering solutions to problems in the aeronautical industry.

## 1.3 Design Freedom

The design freedoms afforded by Rapid Manufacturing are immense and the processes are capable of creating mind boggling geometries. In the short life of layer additive manufacturing technologies the processes have outstripped the capabilities of CAD, in many cases the bottleneck in producing parts is their design, while making them is the easy part. Prior to the advent of these technologies, has mankind ever been in the situation where visualising and designing a product is actually harder than making it?

As these technologies are more commonly used and their products seen by the general public then the creativity that can make full use of the potential of the processes will be realised. As today's computer literate children grow up they will be able to unleash their creativity in ways that had not been possible before. It is possible that three-dimensional modelling packages will be become standard pieces of software – how long before we see Microsoft

CAD? There will need to be considerable work in the development of such packages to suit the new generation of computer literate but non-engineering specialised designers of tomorrow.

## 1.4 Economic for Volumes down to One

The elimination of tooling, for products where machining is difficult or impossible, opens up a host of possibilities for low and medium volume Rapid Manufacture. Indeed, the concept of widespread economic manufacture to a volume of one may, for the first time, be facilitated by this technology. This mode of manufacture is likely to involve the customer very closely and will prompt the need to bring manufacturing close to the point of sale – reversing the current drift of manufacturing from west to east.

The issues brought about by widespread manufacture to a volume of one are many and varied. The process of design is likely to require an increased amount of virtual prototyping with all new products subject to design optimisation and testing using finite element analysis. Production and product proving will be changed and certification standards such as CE etc. will need to be reviewed in the light of these new possibilities. Also the legal ramifications will need to be considered; if a customer designs or has a role in designing his or her own product and a product supplier manufactures the product, who will be liable in the event of a product failure?

## 1.5 Overcoming the Legacy of Rapid Prototyping

In our experience it appears that one of the main stumbling blocks for the increased uptake of Rapid Manufacture is a frequent reticence to accept it as a genuine possibility. For every individual who can see the opportunities that Rapid Manufacturing offers, there is another who prefers to focus on why it can not or will not happen. The latter response is often taken in spite of clear evidence that Rapid Manufacture is already happening.

Rejection of the concept of Rapid Manufacture usually comes in the form of comparison (for example of material properties) with existing processes. The problem at this point is that Rapid Manufacture is seen as merely an extension of Rapid Prototyping and so parts are not seen to be suitable or intended for end use. This 'baggage' of Rapid Prototyping is probably a larger hurdle to the uptake of Rapid Manufacturing than any of the technical issues that we face. Overcoming this viewpoint will take time, more evidence of success and the ability to present the benefits in a clear and balanced way.

## 1.6 A Disruptive Technology

Many of the issues discussed above are clear symptoms of a technology that can be described as disruptive. Rapid Manufacturing offers profound possibilities across a broad spectrum but has initially been met with a wide-ranging degree of acceptance, often leading to lively debate!

## 1.7 A Breakdown of the Field of Rapid Manufacturing

Through our extensive involvement with this new discipline we have identified four key areas of the technology and have arranged the chapters of this book to fall within these areas of:

- Design
- Materials and processes
- Management and organisational issues
- Applications

We are intrigued by what we have found as this technology has developed and are excited about the future that it holds. We hope that this book introduces the topic in a manner that does justice to the next industrial revolution.

# 2

# Unlocking the Design Potential of Rapid Manufacturing

Richard Hague
*Loughborough University*

## 2.1 Introduction

One of the principal advantages of taking an additive (Rapid Manufacturing) approach to manufacturing over more conventional subtractive or formative methods comes not from manufacturing approach per se but from the dramatic advantages that are possible in the area of *design*. This potential for radically different design methodologies is one of the major drivers for the development of Rapid Manufacturing systems and materials and is a powerful reason why some organisations are able to put up with the sometimes severe limitations associated with current Rapid Prototyping (RP) systems to gain an advantage today.

The main benefit to be gained by taking an additive manufacturing approach (including most, but not all, of the currently available RP techniques) is the ability to manufacture parts of virtually any complexity of geometry entirely without the need for tooling. In conventional manufacturing, there is a direct link between the complexity of a part and its cost. In Rapid Manufacturing (RM), not only is complexity independent of cost but also the RM techniques are able to produce virtually any geometry. If this principal were extended to true manufacturing processes then the opportunities for product design and manufacturing are immense.

*Rapid Manufacturing: An Industrial Revolution for the Digital Age*
Editors N. Hopkinson, R.J.M. Hague and P.M. Dickens 

This need for tooling in conventional manufacturing represents one of the most restrictive factors for today's product development. The absence of tooling within the additive manufacturing processes means that many of the restrictions of 'Design for Manufacture and Assembly' (DFMA) [1] that are essential in a modern manufacturing environment are no longer valid [2]. In injection moulding, for example, the need to consider the extraction of the part from the (usually expensive) tool takes an overriding precedence in the design of the part. Thus the high cost and need for tooling greatly limits product design and compromises have to be made. Without the need for tooling or necessity to consider any form of DFMA, the possibilities for design are literally only limited by imagination.

During the last few decades, designers have been educated to develop designs with restricted geometry so that parts can be made easily. The revolutionary aspect of Rapid Manufacturing will be that geometry will no longer be a limiting factor. Compounding the fact that as high volumes do not need to be manufactured to offset the cost of tooling then the possibilities for affordable, highly complex, custom parts become apparent. In theory, each part that is produced could be a custom part and thus there will be the potential to economically 'manufacture to a unit of one' [3]. The ability to produce whatever geometry that is created in a three-dimensional computer aided design (CAD) system actually means that one is entering a new dimension of 'Manufacture for Design' rather than the more conventional 'Design for Manufacture' philosophy [4].

This freedom of design is one of the most important features of RM and is extremely significant for producing parts of complex or customised geometries, which will result in reducing the lead-time and ultimately the overall manufacturing costs for such items. RM will affect manufacturers and customers alike. For manufacturers, costs will be dramatically reduced as no tooling is required and for customers, complex, individualised products will be cost-effectively made that can be configured to personal use, thus giving the potential for much greater customer satisfaction [5].

Rapid Manufacturing will enable fast, flexible and reconfigurable manufacturing to occur that will have enormous benefits to manufacturers and consumers. The elimination of tooling and the subsequent removal of many DFMA criteria will realise significant benefits in the design, manufacture and distribution of a part or components, including:

- Economic low-volume production
- Increased flexibility and productivity
- Design freedom

The subject of 'Design for Manufacture' is potentially broad. However, this chapter will concentrate on the 'freedom of design' aspects and will give details of specific areas of design that are only enabled by taking an additive approach to manufacturing.

## 2.2 Potential of Rapid Manufacturing on Design

The main feature of RM processes is the ability to produce parts of virtually any shape complexity without the need for any tooling. The impact of this factor on the validity of guidelines that designers comply with when they are designing for manufacture and assembly are discussed below.

### *2.2.1 Conventional 'Design for Manufacture' (DFM)*

DFM is a philosophy or mind-set in which manufacturing input is used at the earliest stages of design in order to design parts and products that can be produced more easily and more economically. DFM is any aspect of the design process in which the issues involved in manufacturing the designed object are considered explicitly with a view to influencing the design. Some principals are used for efficient manufacturing, such as: developing a modular design, using standard components and designing for multi-use and to be multi-functional. By far the most important principle is to design for ease of manufacture and fabrication, which could be different depending on the manufacturing processes adopted. These guidelines are well documented elsewhere [1,6].

For years, designers have been restricted in what they can produce as they have generally had to design for manufacture – i.e. adjust their design intent to enable the component (or assembly) to be manufactured using a particular process or processes. In addition, if a mould is used to produce an item, there are therefore automatically inherent restrictions to the design imposed at the very beginning.

As the range of plastic products being produced by RP and RM processes are quite comparable with those of injection moulding of plastics, some of the rules necessary for injection moulding are given here in order to provide a basis for the consideration of design rules for Rapid Manufacturing. These include:

1. *Draft angles.* These are important for ease of removal of parts from moulds. The inclusion of draft angles at the design stage is very important, but often omitted.
2. *Minimising re-entrant features.* 'An easy to manufacture part' must be easily ejected from the mould. Designing undercuts requires the use of side cores. This in turn will require moving parts in the dies that add to the tooling costs considerably. Some parts containing features such as blind

holes and galleries are impossible to manufacture without using very complex and expensive tooling arrangements.
3. *Wall thickness consideration.* Components with thin walls solidify faster, hence reducing warpage and production costs.
4. *Uniform wall thickness.* Non-uniform wall thickness will result in compression and expansion of molecules, resulting in compressive and tensile stresses. The stress in turn will result in cracks, crazing or fractures of moulded parts.
5. *Minimising weld lines.* When different flow fronts (due to obstruction within the mould or various gates) meet each other, this creates weld or fusion lines. These are a source of weakness within the part and should be minimised during design.
6. *Avoiding sharp corners.* These will provide tensile, compressive and shear stress on the moulded parts, which in turn will become stress concentration points, leading to part failure.
7. *Ejection pin marks and gate marks.* These could have an adverse aesthetic effect on the injection-moulded part. However, with adequate consideration their impact could be minimised.
8. *Parting line.* The direction of mould closure and parting line is also crucial in tooling and injected parts. Much consideration and deliberation is needed for their selection.
9. *Minimising sink marks.* These are formed when a thin section becomes solid sooner than a developed thicker section. Sink marks could be less apparent by adequate consideration during design.

### 2.2.2 Conventional Design for Assembly (DFA)

By adopting DFA guidelines at the design stage, significant reductions in manufacturing cost and improvements in the ease of assembly can be achieved [7]. A few of these guidelines are briefly given here [1,6]:

1. *Reducing parts count.* Eliminating unnecessary parts, combining parts or eliminating or reducing the number of fasteners could achieve this.
2. *Reducing handling time.* A few simple, logical and effective rules, such as avoiding tangling and nesting parts or using easy-to-handle symmetrical parts, would result in a more efficient assembly.
3. *Ease of insertion.* This involves designing parts that are easy to align, easy to insert and self-locating with no need to be held in place before insertion of the next part.

### 2.2.3 Impact of RM on DFM and DFA

As the first RM processes will most probably be plastic processing systems, the most immediate competition will be with injection moulding. RM, unlike injection moulding, is a tool-less process, which does not involve any melting and subsequent solidification of materials within the confines of a tool. Therefore, considerations for constant wall thickness (to aid the flow of material), avoidance of sharp corners and minimising weld lines, sink marks, ejection pins, gates marks and draft angles will no longer need to be considered.

However, the significant impact of RM will be on the guidelines associated with minimising complex geometries and features such as undercuts, blind holes, screws, etc. Incorporating such features in conventional injection moulding is not impossible but often requires expensive tooling, extensive tool set-ups, testing runs and prototyping. This inevitably leads to undesirable lead-times and costs. Also, any simple modification in design requires a new set of tooling. However, as RM is a tool-less process, the part complexity is not important and any complex shapes or features produced by CAD can be directly translated into the final product. This is in marked contrast to conventional manufacturing processes.

Also, in injection moulding, the selection of the correct location for the split line – in particular for asymmetrical and complex-shaped components – is quite difficult and is largely dependent on the experience of the tool designer. However, by adopting RM processes and not using any tooling, designers will be entirely freed from this task.

By using RM technologies, it will be possible to reduce the number of parts within an assembly. Therefore, the most important DFA guideline, which concerns the reduction in part count, is easily achievable. In theory it is possible to reduce the number of parts to just one, though in practice this may not feasible as parts are generally not being used in isolation and their interaction with other components would impose limitations on a part's count.

Thus, with the advent of the Rapid Manufacturing techniques, there is the potential for many of the current obstacles to be removed. The following sections discuss the design freedoms afforded by RM and also deal with some potential problems that are likely to occur with the onset of Rapid Manufacturing in general.

## 2.3 Geometrical Freedom

As discussed, one of the major benefits of some additive manufacturing processes is that it is possible to make virtually any complexity of geometry at no extra cost. This is virtually unheard of, as in every conventional manufacturing technique there is a direct link from the cost of a component to the complexity of its design. Therefore, for a given volume of component, it is effectively possible to get the geometry (or complexity) for 'free', as the costs incurred for any given additive manufacturing technique are usually determined by the time to build a certain volume of part, which in turn is determined by the orientation that the component is built in.

Areas of particular interest that are enabled by the freedoms afforded by RM include:

- Design complexity/optimisation
- Parts consolidation
- Body-fitting customisation
- Multiple assemblies manufactured as one

These areas are discussed in greater detail in the following sections.

### *2.3.1 Design Complexity/Optimisation*

The design freedoms afforded by RM will enable increasingly complex designs to be realised that are fully optimised for the function that they are required for. Design optimisation is common in the construction industry where optimal structures for bridges and buildings are derived using optimisation techniques and then subsequently fabricated. For example, Figure 2.1 shows the proposed Beijing National Stadium, which has been designed by Arup for the 2008 Olympics. This building has been designed with a combination of design optimisation and genetic algorithms to produce a truly unique structure, but one that is structurally sound.

It is proposed that, due to the freedoms of design afforded by RM, this approach can be used much more extensively for product design – this approach is less common in the product design arena as the optimised design will often prove impossible to make due to DFM criteria. This is one of the main stumbling blocks for so-called Knowledge-Based Engineering (KBE) systems that often have finite element analysis (FEA) as the kernel.

Initial work at Loughborough University has investigated the use of design optimisation to create complex internal structures. Figure 2.2 illustrates a

**Figure 2.1** Proposed Beijing National Stadium designed by Arup (8)

**Figure 2.2** Conventional front plate

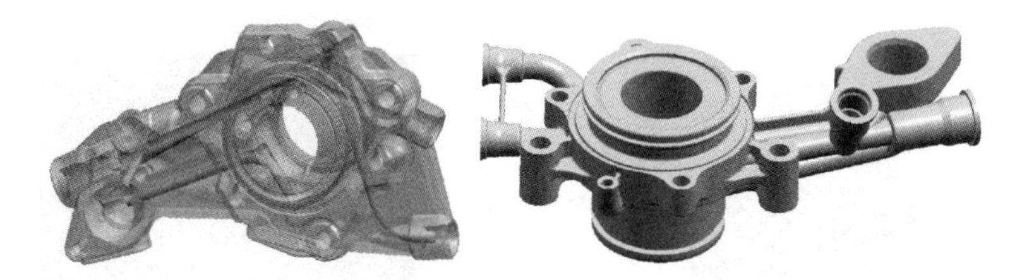

**Figure 2.3** Optimised flow channels and design optimised front plate

diesel front plate manufactured by Delphi Diesel Systems showing internal flow channels that have to be conventionally gun-drilled.

A consideration has been made as to what the design would be like if it were possible to manufacture this by RM. Figure 2.3 shows firstly the flow channels that would be manufacturable by RM followed by a design that is optimised for minimal weight that has been constructed around the flow lines. This approach represents a new philosophy to product design that will have radical implications for the performance of items in the future – only RM allows this.

### *2.3.2 Part Consolidation*

One of the most important opportunities to arise from the ability to 'manufacture for design' comes from the very real potential to consolidate many components into one. This reduction of assemblies has tremendous implications, not just for the actual assembly of the components and the consequent cost savings that can be gained but also from the potential to maximise a design of a product for the purpose in mind and not to have to compromise the design for manufacturing and assembly reasons. Figure 2.4 shows an example of part consolidation where a complex ducting channel assembly has been consolidated into just one part. Such an approach is currently being utilised by US aircraft manufacturers for the production of ducting (on SLS, or selective laser sintering, machines) for use in fighter aircraft.

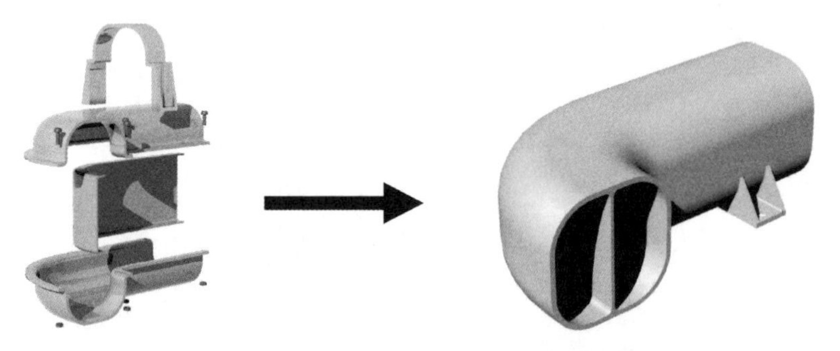

**Figure 2.4** Example of parts consolidation in aircraft ducting (9)

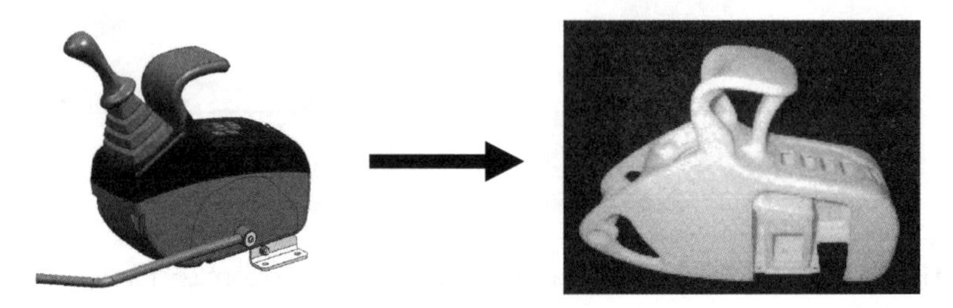

**Figure 2.5** Consolidation of the control pod

Work has been undertaken at Loughborough into the concept of parts consolidation. Figure 2.5 shows how an assembly of over 25 parts has been consolidated into just one piece (with an extra cover) and then manufactured by stereolithography (SL). Early calculations suggest that, not withstanding the material properties, the SL pod would be commercially viable due to the low number that would be manufactured and also the advantages that are gained by consolidating the assembly into just one part.

Another example of parts consolidation from Loughborough University involves the redesign and manufacture of car door handles with Jaguar Cars. An initial assembly comprising eleven components made from eight different materials was simplified to a single component made from a single material that was manufactured by SLS. In this case the work showed how the design freedoms afforded by RM could ease the end-of-life recycling.

### *2.3.3 Body Fitting Customisation*

The production of body fit parts and core customisation is not novel. However, the creation of customised products using conventional skills and technologies, especially truly body fitting customised products, has traditionally been very labour-intensive and essentially craft-based. Thus, partly due to the costs of labour, customised products are usually out of reach of the general public, who are forced to buy mass-produced goods.

As true customisation is at present not feasible for the mass market, the concept of 'mass customisation' is currently employed to give some degree of customisation. However, mass customisation is really achieved by 'modularisation' – the production of modules that can be bolted together in varying configurations, which that gives the economies of mass production but allows some choice in the product. By using management techniques such as postponement, the decision of how the final product is configured can be delayed to allow for greater degrees of customisation. However, standard manufacturing techniques are still employed (for the modules) and therefore there is still the need for costly tooling.

However, through the adoption of reverse engineering and Rapid Manufacturing, the era of cost-effective customisation for the masses is not far off. With the advent of RM the production method and processes involved for customised parts would not change from part to part. Thus, the economic argument for providing core (body fitting) customisation is greatly enhanced. This approach is already commercial reality as Siemens and Phonak are using laser sintering and stereolithography to manufacture bespoke hearing aids for end use (see Chapter 12).

At Loughborough University, this approach is being further considered by the MANRM project (under a DTI Foresight Vehicle Initiative) where a methodology has been established for the capture of the correct deformed geometry that is required for creating the body fitting customised parts. This project is particularly focused on the aerospace/automotive sectors. Figure 2.6 shows a body fitting seat that has been produced for one of the project partners, MG Rover. The parts are then manufactured automatically using RM technologies that require no expensive tooling. In the near future, this work will lead to affordable custom fitting products for the general public.

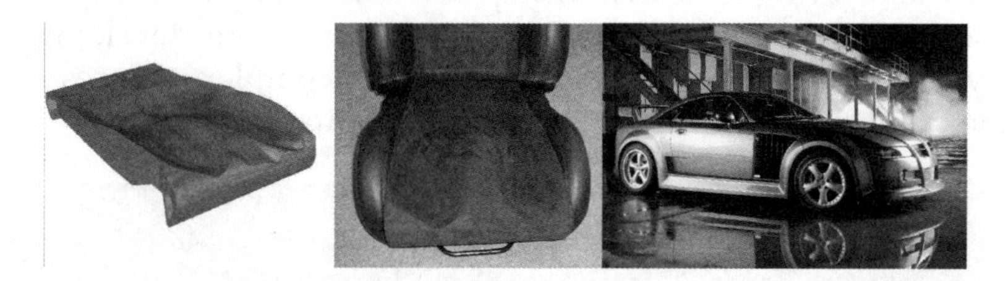

**Figure 2.6** Body fitting seating platform and seat for the MG SV. (Reproduced with permission of MG Rover Group)

### 2.3.4 Multiple Assemblies: Textiles

An entirely novel area that has received virtually no consideration for RM, but one that has vast and exciting potential for future applications is that of smart textiles. Conventional sheet textiles, as with any other product, have to be constructed with the manufacturing process in mind and thus textiles are

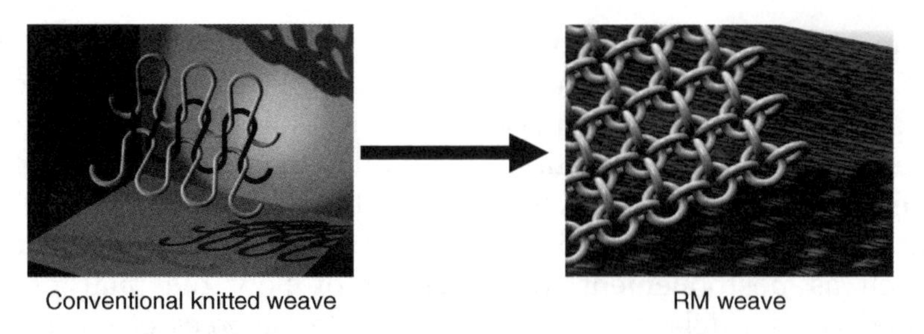


**Figure 2.7** Comparison of conventional versus RM textiles

fundamentally limited by the need to design for manufacture. Conventional fabric/textile construction uses centuries-old principles – the key for RM fabrics (at present) is to move from continuous fibres to individual links. This was first proposed by Jiri Evenhuis and Janne Kyttanen [10] in 1999. This is demonstrated in Figure 2.7.

However, there are many research issues that need to be overcome before RM produced garments are a reality. These research issues include:

- Link design
- Generation of three-dimensional data
- Lofting of data over conformal surfaces
- Very large data sets
- Collapsing of the structures for efficient manufacture
- Rapid Manufacturing process resolution

One of the most fundamental issues for creating body fitting RM textiles is the difficulty of using current CAD systems, which are not intended or capable of creating such complex assemblies. However, the Loughborough Rapid Manufacturing Research Group (in collaboration with Nottingham University Composite Materials Research Group) have developed a preliminary methodology to wrap textile links over complex surfaces and then manufacture them by SLS. This is shown in Figure 2.8.

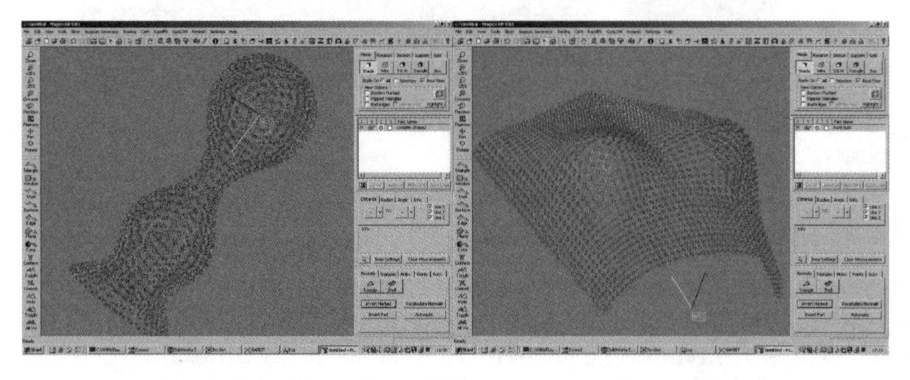

**Figure 2.8** Wrapping of links over complex surfaces

**Figure 2.9** Structure collapsing for efficient manufacture

The current manufacture of all RM textile sheets, garments or products requires the design and CAD data to directly mimic the finished component in its three-dimensional form (as shown in Figure 2.8). Therefore, in order to fit the garment in the build envelope and also limit the $z$ height or number of layers (and thus minimise the cost), it is necessary to then collapse these data into a manufacturable form. Initial work has established a methodology for performing this for simple link constructions (as demonstrated in Figure 2.9), but much work is required to extend this to more complex constructions.

In summary, it is postulated that future RM systems will be capable of creating textiles directly and will be extended to the automated generation of the items that they are assembled into (e.g. clothing) using RM as the enabling technology. As a demonstration, Figure 2.10 shows an example of an RM dress that has been produced using current laser sintering technologies, which is the world's first fully conformal (body fitting) textile

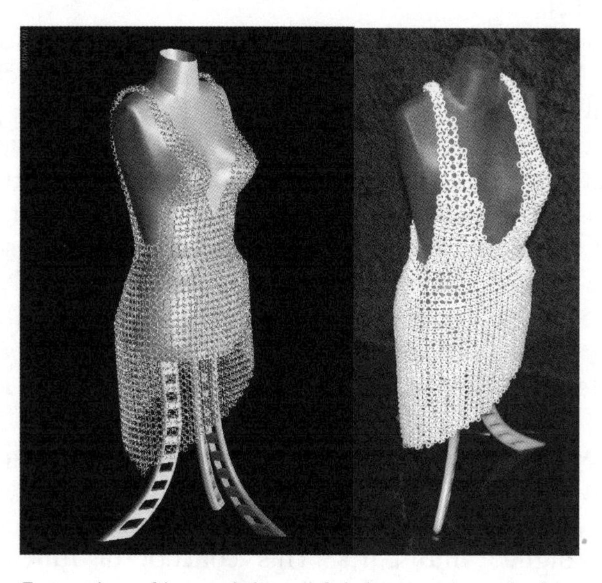

**Figure 2.10** Examples of laser sintered fabrics produced at Loughborough

garment to be produced directly in its assembled state. This, again, is only achievable by utilising an additive manufacturing approach.

Work is now underway aimed at investigating the potential for micro-level design and manufacture in the context of RM and initiating research into the automated knowledge-based generation of design optimised smart textiles producible via RM. In the future, there is much scope for:

- Seamless garments that can be manufactured fully assembled
- Variable 'weave'
- Products that transition from a solid to a textile configuration, thus giving possibilities for optimised footwear
- Smart textiles with built-in functionality (e.g. truly wearable computers, built-in chemical/biological detection, custom fitting armoured jackets, etc.)

## 2.4 Material Combinations

When objects are formed in moulds, they are generally formed in one homogeneous material. Even in the case of an over-moulded component, where there can be two or more homogeneous materials in one finished part, there is a definitive boundary between one material and the other. In the future, with some of the additive manufacturing processes there is the potential to mix and grade materials in any combination that is desired, thus enabling materials with certain properties to be deposited where they are needed [11,12].

The over-moulding technique is a classic example of how design can be influenced by the availability of a manufacturing technique. Over-moulding allows designers, within limits, the ability to produce parts that have added functionality and enhanced design. Indeed, the design of over-moulded components very often incorporates different material combinations to accentuate the design to the extent that designers are able to exploit the delineation of the different materials used to produce design features as well as extra functionality. This is perfectly illustrated by the simple case of a toothbrush – an everyday item that will often include over-moulding to give a handle that is stiff, with an over-moulded grip and a different material at the neck to give a flexible head. This is illustrated in Figure 2.11.

Given that RM potentially allows the development of multiple materials to be deposited in any location or combination that the designer requires, this has potentially enormous implications for the functionality and aesthetics that can be designed into parts. This concept of functionally graded materials (FGMs) is further discussed in Chapter 7.

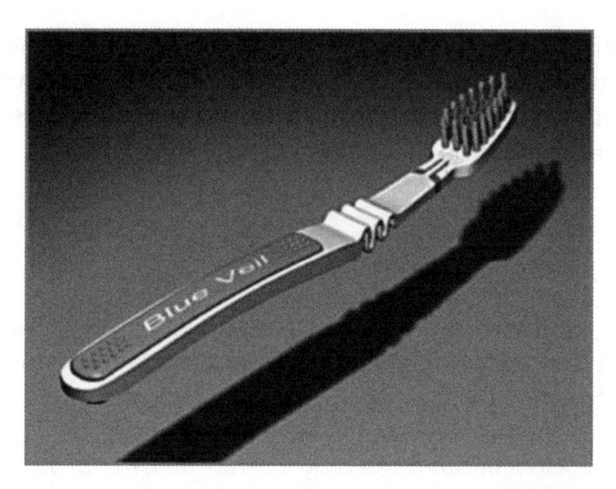


**Figure 2.11** Example of how over-moulding improves design and functionality

## 2.5 Summary

The possibilities offered by Rapid Manufacturing are enormous. Suddenly, designers will be able to manufacture almost any shape that they come up with and will no longer be constrained by the necessity to produce parts in moulds. In addition, using processes such as the laser sintering of dissimilar powders, RM will provide designers with new and exotic materials not available to other manufacturing processes.

At the design phase, RM allows almost whatever shape is desired as the mould process will no longer limit design. This means objects can be designed with re-entrant features, no draft angles, unlimited wall thickness and increased complexity, with none of the limitations imposed by either the moulding process or the tool making process, as neither will be required.

One of the most profound implications of RM on design will be that, without the cost of tooling to amortise into the parts produced, each component can be different, potentially allowing for true mass customisation of each and every product. With developments in web-enabled software, high levels of computer literacy and Internet connectivity in the home, the technologies are not far from giving the consumer the ability to modify the design of the product they desire for themselves. Although some way off, it is conceivable that the consumer may – for a price – want to influence the design of their new sunglasses, mobile phone casing, steering wheel grip, surgical instrument, prosthetic part or favourite kitchen utensil, etc., and then send the data back to the manufacturer to be made for them.

RM will become more of a reality when the properties of the materials that are produced become more acceptable and consistent. This materials research is one of the main stumbling blocks to the adoption of these additive manufacturing techniques for end-use parts and is the subject of much current research. However, many organisations are willing to accept the materials limitations that are in evidence today to gain an advantage from the design possibilities.

## References

1. Boothroyd, G., Dewhurst, P. and Knight, W. (1994) *Product Design for Manufacture and Assembly*, Marcel Dekker Inc., New York.
2. Mansour, S. and Hague, R. (2003) Impact of rapid manufacturing on design for manufacture for injection moulding, *Proceedings of the Institution of Mechanical Engineers, Part B: Journal of Engineering Manufacture*, **217**(B4), 453–61.
3. UK manufacturing: we can make it better, Foresight Manufacturing 2020 Panel, Final Report, Findlay Publications, p. 12.
4. Campbell, R.I., Hague, R.J.M., Sener, B. and Wormald, P.W. (2003) The potential for the bespoke industrial designer, *The Design Journal*, **6**(3), 24–34.
5. Hague, R.J., Campbell, R.I. and Dickens, P.M. (2003) Implications on design of rapid manufacturing, *Proceedings of the Institution of Mechanical Engineers, Part C: Journal of Mechanical Engineering Science*, **217**(C1), 25–30.
6. Poli, C. (2001) *Design for Manufacturing, A Structured Approach*, Butterworth- Heinemann, Boston, Massachusetts.
7. Fox, S., Marsh, L. and Cockerham, G. (2001) Design for manufacture: a strategy for successful application to building, *Construction Management and Economics*, **19**, 493–502.
8. www.arup.com
9. www.3dsystems.com
10. www.freedomofcreation.com
11. Anon (2001) The solid future of rapid prototyping, *The Economist Technology Quarterly*, 24 March 2001, pp. 47–9.
12. Jacobs, P.F. (2002) From stereolithography to LENS: a brief history of laser fabrication, International Conference on *Metal Powder Deposition for Rapid Manufacturing*, San Antonio, Texas, 8–10 April 2002.

# 3

# Customer Input and Customisation

R.I. Campbell
*Loughborough University*

## 3.1 Introduction

This chapter explores how Rapid Manufacturing (RM) will enable a push towards affordable customised products which will necessitate increased customer input to the design process. The term 'customer' is normally understood to mean the person who buys and then uses a product. However, the issue is somewhat clouded when the person who buys a product and the person who uses it are not the same individual. For example, the decision to purchase a particular vehicle to be used as a rental car is made by a company employee who may never actually drive it. One person is the customer whereas someone else is the user. Another possibility is that a product is purchased by an initial customer who then sells it on to someone else. For example, the owner of a small shop is the customer who buys from a soft-drinks manufacturer but the drink is then purchased and consumed by one of the shop's clientele. To avoid this possible confusion, some companies will use the terms end user or consumer to distinguish between the person who buys the product (the customer) and the person who uses it.

However, in this chapter, the term customer will be used in a more general way to refer to the person who will be the final owner, or at least user, of the product. This is the person (or persons) whose input to the product design is most valuable and therefore should be most sought after. The chapter

*Rapid Manufacturing: An Industrial Revolution for the Digital Age*
Editors N. Hopkinson, R.J.M. Hague and P.M. Dickens 

examines why customer input is required, the diverse nature of this input, how it can be captured within the design process and how it can be used to enable RM to produce customised products.

## 3.2 Why Is Customer Input Needed?

Many designers may be tempted to take the view that they know what and how to design and they do not need any outside influence to distract them from achieving their aims. This is a particularly attractive position if the added time and cost of customer input is taken into consideration. A further obstacle is the often present requirement to keep a product confidential before it is launched. Exposure of the product concept to potential customers could lead to confidentiality being compromised. These issues and others undoubtedly mean that there are time, cost and risk penalties associated with involving customers in the design process.

In the past, this has led to some companies relegating customer input to an acceptability study of the final design to see if any 'minor tweaking' is necessary. Perhaps one of the best known examples of this is the Sinclair C5 electric vehicle shown in Figure 3.1. It was the brainchild of one man who thought he knew what the market required and no doubt had market survey data that proved this. The idea of an environmentally friendly, next generation electric vehicle that would revolutionise urban transport was heralded in the press long before the product launch, but the public were not allowed to see what it would actually look like. When it was finally released to the public view, it quickly became an object of derision as it fell far short of what potential customers had been expecting. Although C5 sales got off to a quick start, they soon dwindled and production was discontinued [1]. It soon had

**Figure 3.1** Sinclair C5 electric vehicle (2). (Reproduced with permission of Mr P. Andrews, retro-trader.com)

to have its price tag drastically reduced and many C5s ended up being given away as raffle prizes, promotional items, etc. The previously revered figure of Clive Sinclair became somewhat an object of humour. Had customers been shown what was on offer at an early stage of the design process, it is likely that a radical rethink would have resulted and perhaps the project would have been shelved altogether.

There are many such examples of product failures that could have been avoided if customer input had been treated with a higher degree of importance during design. Indeed, it could be argued that without customer input, a product will succeed merely as a matter of chance rather than through any methodical process. There are also well-known examples of this phenomenon, e.g. the original Mini car. It was also the brainchild of one man who was adamant that his ideas and his alone would be embodied in the design. This is supposedly why the original design did not have an interior clock! However, this type of 'lone-designer' success should be seen as rare and in today's highly competitive global market, companies cannot afford to take such a high risk as failure of one major product launch could lead to financial ruin.

Companies need to bring customers into the design process at every stage, first of all capturing their requirements and then verifying that successive design iterations continue to meet these. A good example of this is when Ford Motor Company wanted to develop a replacement for their highly successful European car, the Escort. They made extensive use of customer focus groups throughout the project and the end result was a vehicle (appropriately named the Focus) that was voted European Car of the Year 1999 and has gone on to be a highly successful car around the world. No amount of customer input can give a 100% guarantee of product success but it can certainly help to minimise the risk of failure. As customers become increasingly discerning and demanding, so this input is becoming even more critical. Nowhere is this more so than in the realm of customised or individualised products, for which customer input can never be treated as an option. Since RM can take customisation to a new level, high-quality customer input is needed more than ever.

## 3.3 What Input Can the Customer Make?

The aim of any customer input exercise should be to capture their requirements into the product design specification (PDS). Such requirements can range from 'hard' functionality such as power output or load-bearing capability to much 'softer' qualities like 'I want this product to make me feel successful'. Some requirements might be shared by a large number of people in society, e.g. 'must be safe to operate', whereas others may be specific to an individual, e.g. 'must fit my foot'. Obviously, the wide diversity of requirements that can be captured will require a range of different tools and techniques. The problem is that it is the more difficult-to-capture qualities that will often have the most impact upon product success. The main categories of input are discussed below.

### *3.3.1 Functional Requirements*

These refer to what the product will actually do, i.e. its primary purpose. This will include immediate performance targets but also longer term aspects such as reliability, serviceability and the life in service that customers will expect.

### *3.3.2 Environmental Requirements*

These concern the impact that the manufacture, use and disposal of the product will have upon the environment. Customers may have a view on manufacturing processes or materials that they regard as 'environmentally unfriendly'. They would not buy a product made in this way. Likewise, they may only wish to own a product that has lower energy consumption and that is easy to recycle.

### *3.3.3 Ergonomic Requirements*

These are best understood as the product features that lead to ease-of-use. Therefore, they include all aspects of the human–product interface, e.g. the size of a handle, comfort of seat cushions, overall weight, font used in decal lettering, clarity of instruction booklets, etc. In terms of the physical size aspect, ergonomists often make use of standard anthropometric data that give various human dimensions in terms of percentile values (see Figure 3.2 for an example). These can be used to estimate what percentage of customers will be accommodated by the product.

### *3.3.4 User-fit Requirements*

This could possibly be seen as a subset of ergonomic requirements. It refers to the physical interface between the product and the customer's body. However, it has the extra connotation that the product must be fitted to the individual, not merely a percentile range. Example products where this is crucial are dentures, spectacles and orthopaedic devices.

### *3.3.5 Aesthetic Requirements*

The term aesthetics refers to the impact that a product will have upon the human senses, most notably visual impact. This will include not only the

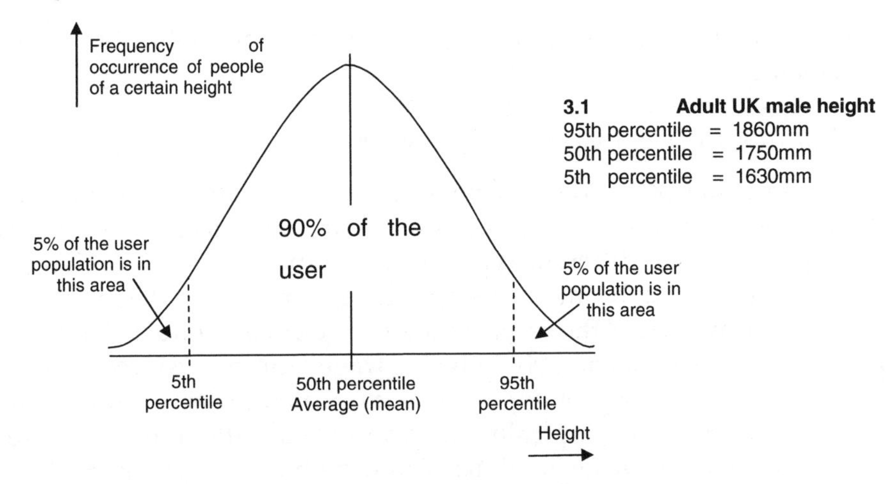


**Figure 3.2** Anthropometric data for adult UK male height

overall form of the product but also the colour scheme and surface textures. Other important sensual requirements may include how the product feels to touch and how it sounds.

### *3.3.6 Emotional Requirements*

This is perhaps the most difficult category to define because it may encompass all of the above. For example, think of the various requirements that would be necessary for a product to provide a feeling of luxury. However, the unique aspect of emotional requirements is that they are normally not directly measurable in the product. Rather, it is the effect that the product creates within the customer. This also tends to be the most difficult area to predict and to measure.

## 3.4 How Can Customer Input Be Captured?

This will obviously depend heavily upon the nature of the input. In most cases, the potential customers are exposed to an existing product design or a new design concept and asked for their opinions. It is preferable that the customers be allowed to actually use the product (or prototype) and even more desirable if they could do so in a normal operating environment. Various aspects of the design will be evaluated and sometimes rated against some sort of benchmark, for example a competitor's product. Questionnaires can be used to assist with this. The results from many users will be compiled, analysed and then translated into specific requirements (often the most problematic aspect of the process). Techniques available to help with this process are well covered elsewhere [3,4]. This chapter concentrates on the role that RM and related technologies can play in capturing customer requirements.

### *3.4.1 Rapid Manufacturing of Prototypes*

Rapid prototyping (RP) models have brought a new dimension to capturing customer requirements. Prior to the use of RP, designers would normally have to produce two different types of prototype for a new design [5]. A 'block model' would be made using manual modelling techniques that would closely resemble the appearance of the final product. However, it would be non-functional and could be too fragile for the customer to handle. Additionally, a working prototype would be created to demonstrate how the product would function. Typically, this would bear little resemblance to the final product. Customer evaluations would be hindered by the fact that it was impossible to get a holistic view of the product from either prototype. This would severely limit the quality of the opinions produced. Using RP it is possible to create accurate geometric models directly from computer aided design (CAD) data which will include all the internal details. Using secondary processing such as investment casting, these models can be converted into functional components and assembled into a fully working 'appearance prototype' (see Figure 3.3 for an example). RM will take this a stage further since fully functional prototypes made in the final production material will be produced. This enables potential customers to use the prototype as they would the final product, i.e. in the normal environment and without fear of damaging it. This will yield a more representative set of opinions compared to previous methods.

**Figure 3.3** Example of appearance prototype produced using RP

### *3.4.2 Reverse Engineering*

For the user-fit type of requirement, it will be necessary to capture shape data relating to the customer. For example, if the requirement for a hearing aid is to fit an individual's ear precisely, then there needs to be a method for capturing this shape. One such technique is reverse engineering. This is where some kind of probe or sensing device is used to capture Cartesian coordinate data describing the three-dimensional shape of an object. Many methods are available including touch probes, optical cameras and laser scanning. A listing of other reverse engineering techniques is given by Wohlers [6].

Reverse engineering can be used to capture data directly from the human body, but there are problems with keeping the subject stationary, with predicting how the body shape will deform on contact with the product and, in certain cases, problems associated with the need for the subject's privacy. As a result, it is more common to use a casting technique to create an inverse mould of the target body part. This mould can then be scanned or digitised over a prolonged period. The three-dimensional data, in the form of a 'point cloud', is used as the basis of a CAD model for the user-fit interface of the product (see Figure 3.4). The CAD design can then be produced as an RP model and presented to the customer for further feedback. If it is deemed satisfactory, the final product can then be manufactured using further CAD/CAM (computer aided manufacture) techniques.

### *3.4.3 Interactive CAD Models*

This is a relatively new and unexplored aspect of capturing customer input that could prove to be very valuable. The basic idea is that the customer sits down in front of a CAD screen with the designer. An existing product design is presented to the customer and he or she is asked to make suggestions for its improvement. The designer then uses the parametric nature of the CAD model to make rapid on-screen modifications until the

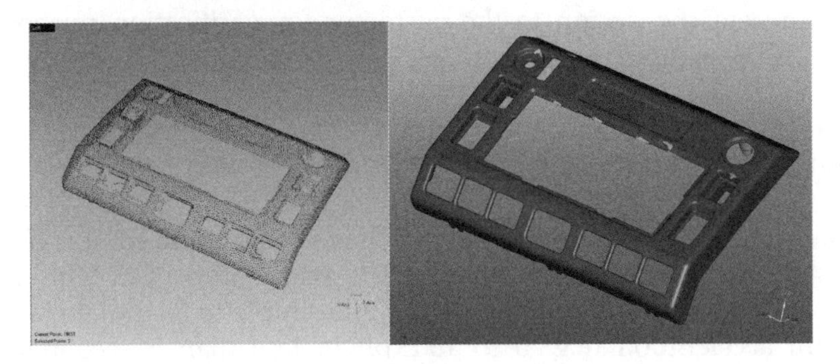

**Figure 3.4** Point cloud (left) used as the basis of a CAD model (right)

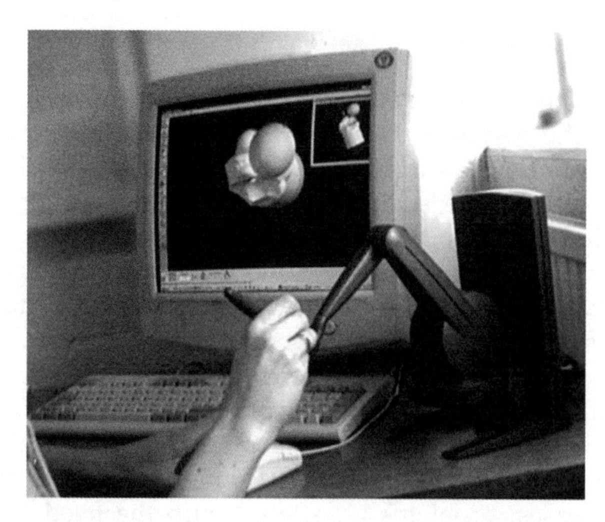

**Figure 3.5** Virtual sculpting system showing a hand-held haptic interface

design meets the customer's requirements. This is particularly well suited to aesthetic requirements. The main drawback at present is that most CAD systems are too inflexible at making changes and the time needed for regenerating a complex parametric model is excessive. However, some interesting work has been conducted at Loughborough University using 'virtual sculpting' CAD techniques where changes can be totally freeform in nature and are immediately executed (see Figure 3.5).

The intuitive haptic interface used means that the customers themselves could even attempt to make changes to the design. It should be noted that such interactive sessions are not aimed at developing the final product design but rather at gaining a better insight into what the customer wants. This information is then used by the designer to help shape the next iteration of the design. A further development of this technique is to allow the customer to 'design' the approximate product shape they require in a physical medium, and then to use reverse engineering to capture this as a CAD model that can be refined by the designer [7]. Once again, RM can be used to present the design to the customer for verification purposes. An example of this is given in section 3.9.

## 3.5 Using Customer Input within the Design Process

Once the various customer requirements have been captured, it is necessary to translate these into measurable targets within the PDS. This is more problematic for the softer qualities. It is quite simple to formulate targets for maximum acceleration (e.g. 0 to 100 kph in less than 5 seconds) or the number of passengers to be carried (e.g. four adults or two adults and three children) but it is not so easy to define a target for 'looks robust'. However, even in these more ethereal situations, targets can be set (e.g. at least 80% of the target user group must judge the appearance to be more robust than the competitor's product X). The upshot of this is that every kind of requirement which will have a bearing upon whether or not the customer would want to own the product can be (and should be) captured and then converted into design parameters.

This is not a trivial task since much of the data will be qualitative and not directly related to design parameters. However, some lessons can be learnt from how ergonomic data has been successfully incorporated into the design process [8]. This research showed that failure to make proper use of customer-derived data was almost always due to poor communication between different personnel in the design process. This can be partly avoided by giving designers direct contact with the customer to obtain the data. However, in larger companies, it will not always be the same designer who will go on to work with the data. Therefore, a paper-based method known as an ergonomics checklist has been developed to encourage designers to use customer data within the design process. Originally developed for ergonomic requirements, it could be expanded to include other aspects also. The checklist provides the designer with a list of ergonomics issues that may have an influence upon the customer acceptance of the product being designed. The aim is for the designer to identify the ergonomic issues and the design implications during the early stages of the design process. These could then be discussed with an ergonomist to ensure that the ergonomics requirements were incorporated into the overall design of the product.

This method has proved to be particularly effective when training designers. An extended 'checklist' could be developed to incorporate requirements of other types that have been identified as key to the customer. Having used the checklist for some time, it would be expected that each designer would begin to take customer requirements into consideration as a matter of course.

Other attempts have been made at developing a procedure for closer involvement of customers within the design process. For example, Khalid [9] has proposed a four-step iterative approach as follows:

1. *Survey*. Identify user needs through a questionnaire.
2. *Focus groups*. Measure responses to significant needs using focus groups.
3. *Design*. Integrate user needs within the conceptual design.
4. *Simulation*. Evaluate the intended use and purchase of the product using end users.

This approach is aimed particularly at obtaining effective responses and envisages making extensive use of the World Wide Web. Such use of the

Web to obtain a customer reaction to design concepts has been explored. However, several other methods are available such as product-in-use, customer clinics, customer self-report and designer/customer workshops [10]. The key issue still to be researched is knowing in what ways to involve customers to elicit the most valuable information.

## 3.6 What Is Customisation?

In general, customisation can be thought of as the process of taking a general product design concept and tailoring it to the needs of a specific customer. However, there are different types of customisation and, consequently, different methods of achieving it. At one extreme there is the notion of producing a bespoke product that has been designed from its outset with one customer in mind and aims to satisfy the requirements of that individual and no others. An example of this could be a uniquely commissioned piece of jewellery that will never be replicated. At the other extreme is the modification of one feature of an otherwise standard product. For example, changing its colour or having one component that can be ordered in different sizes. Here, the colour or size could be selected from an infinite variety to suit just one person or it could be selected from a range of options that would be likely to satisfy a group of customers. Somewhere between these two extremes will be the method of modularisation where a highly customised product can be generated by selecting from several ranges of available options. This is a method typically used in the automotive industry where the offer of several alternatives for many features yields a very high number of product configurations. In theory, several thousand cars could roll off the assembly line before two were produced that were exactly the same. The relationship between the type of customisation and number and range of available options is shown in Figure 3.6.

RM lends itself to all types of customisation but is perhaps best suited to creating an infinite number of choices for one or more features. The customisation of the product could be in terms of different functionality, different aesthetics, different user-fit, etc. Every product made using RM would be unique in some way. This effectively means that manufacturing would be on a 'batch size of one' basis, something that RM can handle much more easily than conventional manufacturing processes. This does not necessarily imply that products would be made one at a time, but rather that many variations of the product could be made simultaneously. The geometry (and perhaps material characteristics) of the product's components would be changed in response to customer requirements. If some of the components were not to be customised, e.g. for safety reasons, these could be made in larger batches and assembled with the RM components to

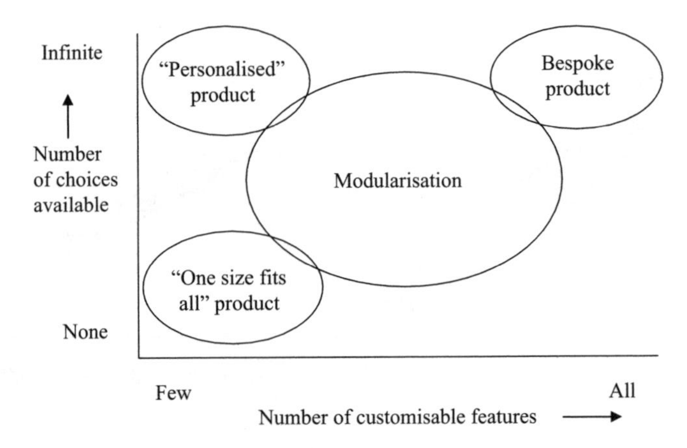


**Figure 3.6** Different types of customisation

produce the final product. Product value would be added by the fact that customers will often pay a premium price for a product that they know is uniquely theirs.

## 3.7 Determining Which Features to Customise

RM could be used to customise virtually every feature of a product but this would seldom be desirable. The main reason behind customisation is to add value to a product. There is very little point in customising a product feature that will not add value. Therefore, before customisation strategies are chosen, a function analysis of the product design should be undertaken. Function analysis attempts to determine the relative contribution of each feature in a product towards its overall value. For example, the value of a toothbrush will come from both the functionality of the handle, stem and bristles and also its aesthetic appearance. By showing potential customers two alternative versions of the design (one with a very simple, box-like shape and the other with a highly developed organic form, as shown in Figure 3.7), and asking them how much they would be willing to pay for each, the value contribution of the aesthetic appearance can be approximated. Obviously, for more complex products with many more features, the functional analysis will become much more complicated. The outcome for the functional analysis will be a list of possible product features together with their relative values. These should add up to the price that the product can be sold for.

For a product that is to be customised, the function analysis must take a slightly different form. Rather than just assuming that a feature may or may not be present, there is now the possibility that the feature can be customised precisely to an individual's requirements. Going back to the toothbrush

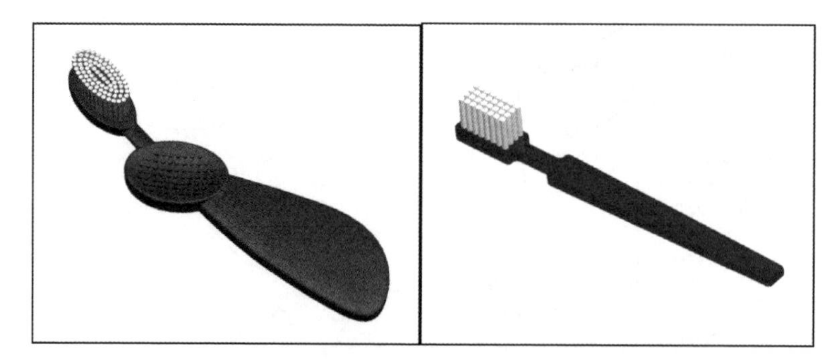

**Figure 3.7** Two versions of a product with radically different aesthetics

example, a customisable feature might be a handle that fits exactly to the customer's grip. An estimation must be made as to how much extra a customer would pay for this specific customised feature. The outcome from this extended function analysis would be a list of features together with the extra price that customers would pay for these to be customised. The design team must now estimate how much it would cost to customise each feature using RM. The cost of customisation will include the extra design effort, higher manufacturing cost (e.g. from using RM) and the increased cost of product support (e.g. carrying more versions of spare parts). The value index for each feature can then be calculated from

$$\text{Value index} = \frac{\text{Extra price paid by customer}}{\text{Cost of customisation}}$$

Features with the highest value indices should be selected for customisation as these will give the greatest return on investment.

Some features will not be suitable for customisation, the most notable being safety critical features. If there are features that will have an impact upon the safe operation of the product then it will not be practical to vary these. If they were altered, safety tests would have to be conducted for every version of the product. This would normally be prohibitively expensive and would delay delivery of the customised product. Other features that might not be suitable for customisation are those that form an important part of the brand identity. It is unlikely that Rolls Royce would want to see their Spirit of Ecstasy replaced by Mickey Mouse on a car owned by a Disney executive.

## 3.8 Additional Customisation Issues

Once the features to be customised have been selected, it might seem a rather straightforward task to have the required parts made through RM and then assembled into the product. However, apart from the technical issue of which RM process to use, there are other issues that arise. These include the location of the customisation process, ownership of intellectual property, product liability and product resale. Most of these issues are covered elsewhere in this book but product resale is of particular concern to the customer and is discussed below.

If a customer purchased a product with a significant number of customised features, it could present a problem if the product was sold on to a new owner. It might be that the new owner has no desire to own these customised features or that the original owner wants to keep them for their own exclusive use. In either case, a solution must be found. The manufacturer or a third party could provide an exchange service where one set of customised parts are exchanged for a new set. This could be done as part of a 'trade-in deal' where some of the original parts would then be transferred on to the new product. For example, if a customer has a customised driver's seat on his present car, this could be removed and fitted to the new car he is about to purchase. A standard seat or a different customised seat would then be fitted to the old car before selling it on to the next owner. A related problem that is not so easily solved is when a product is used concurrently by several people. Any feature that has been customised for one user may not be suitable for the others, e.g. the driving seat in a car driven by several members of one family. It would not be practical to continually swap the customised features.

Product resale and other important issues make it essential that customisation is undertaken as an integrated part of a company's product strategy. Any sort of ad hoc or piecemeal attitude to customisation could land a manufacturer in an organisational and legal mess. Amalgamating RM into an existing business process may not be sufficient. A wholesale re-organisation of production and support systems could become necessary, particularly for more complex products.

## 3.9 Case Study – Customising Garden Fork Handles

As a demonstration of how customer input can be captured and used within the design process to create customised products, a simple case study was undertaken. The product chosen was a small, hand-held gardening fork (the original design is shown in Figure 3.8). The product was selected because it embodied several of the requirement types listed in Section 3.3. The aim was to create customised handles for four individual users according to their specific requirements. The following process was used:

**Figure 3.8** Original gardening fork design

1. Discuss handle requirements through a semi-structured interview.
2. Evaluate the original handle design against a set of predetermined criteria such as grip, aesthetics, usability, etc.
3. Generate an improved user-fit design, recording ideas in verbal, sketch and written format.
4. Capture user-fit and other ergonomic requirements using modelling clay.
5. Translate into a CAD model (using reverse engineering if necessary).
6. Capture and verify aesthetic requirements using CAD rendering.
7. Verify functional requirements using an RP model.

The modelling and prototyping stages are now discussed in turn.

### *3.9.1 Customer Input Through the Use of Modelling Clay*

Customers were shown the original fork design and then provided with the metal element (shaft and prongs) together with air-drying modelling clay. They were asked to model their own design of handle that would fit their hand as they desired, include other ergonomic aspects such as finger grips or wrist supports and functional aspects such as hanging holes. They were encouraged to attach the clay to the metal element during this process to give a representative feel of weight and balance. An image of one of the new designs being modelled is shown in Figure 3.9.

### *3.9.2 Translation into a CAD Model*

Some of the handle designs created were relatively simple in shape and it was possible to model them in CAD through direct observation. However, some of them were more complex and reverse engineering had to be used. Three-dimensional laser scanning was undertaken with a three-dimensional Scanners' ModelMaker and a FARO arm system (see Figure 3.10). The point

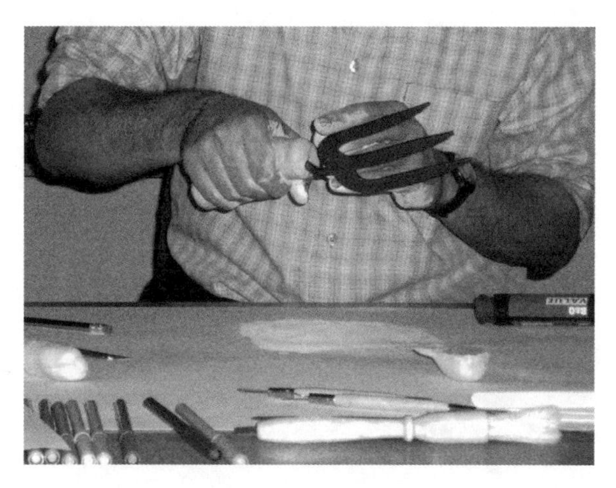

**Figure 3.9** One of the new handle designs being modelled

clouds of data were imported into Geomagics Studio software where it was merged and refined before being used to create non-uniform rational B-spline (NURB) surfaces. If necessary, these surfaces were further refined using the freeform virtual sculpting system and were then used as the basis for building a solid CAD model within Solidworks.

### *3.9.3 CAD Rendering*

Once the CAD model had been completed, it could then be used as the basis for high-quality rendered images that were used to convey alternative colours and surface textures that the finished product could have. The users were shown different versions of these until they were happy with the

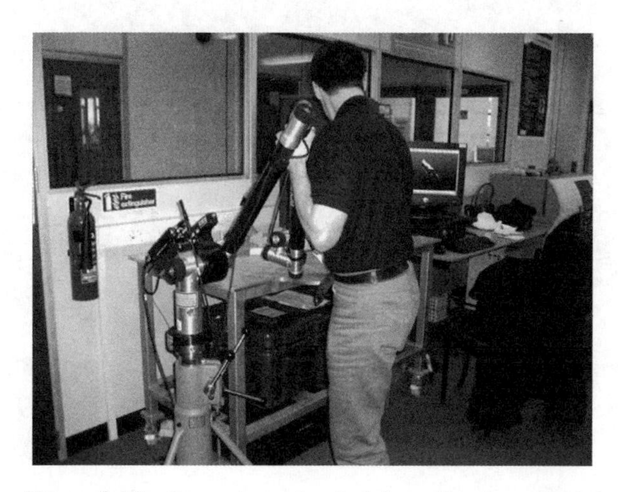

**Figure 3.10** Scanning of a fork handle clay model

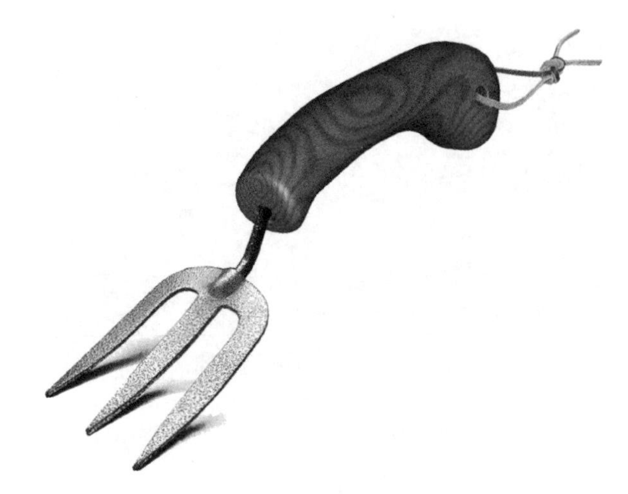

**Figure 3.11** CAD rendering showing wooden handled fork

aesthetic appearance of the design. The example shown in Figure 3.11 is a representation of what a wooden handled fork would look like.

### *3.9.4 Verification of Functionality*

The final stage in the process was to verify that the new design of handle met the customer's requirements for functionality. This was achieved by creating an RP model of the handle using an FDM 2000 machine with ABS material. The handle was hand finished and attached to the metal element of the fork ready for functional testing (see Figure 3.12).

**Figure 3.12** Verification of fork handle functionality

The outcome of this case study was proof that customers can become closely involved in the design process. Designers can work directly with them to capture requirements that can then be used to produce new product designs. The case study also demonstrated the role that RM and related technologies can play in this interaction. A simple product was deliberately chosen for the case study but the principles followed are applicable to more complex products also.

## 3.10 Conclusions

Customers are becoming much more discerning and selective, so for any product to be successful, it must be highly attractive to the customer. Therefore, designers are required who have specific skills to analyse customer wishes, needs and requirements and understand how these can be translated to an end experience for the customer. Designers need to 'get inside the heads' of customers by observing them using the product in the real world. They need to increase their insight into the lives of potential customers. This will help expose them to the often unspoken needs and attitudes of users. This already happens in many smaller, innovation-led companies but is quite rare in larger, more organised companies [11]. The aim is to identify the aspects of proposed designs that will delight the customer. Designers will need tools to help them learn what creates a pleasurable experience for users and not simply rely on prioritising product functions. Likewise, the common practice of designers receiving customer preferences via a separate marketing or market research department is no longer sufficient. Designers must be trained to communicate directly with customers. A term that has been coined to describe such a person is a 'user-centred designer'.

In addition, a new way of designing that would always start and end with customers and be focused on them throughout is required. Such a process would follow a number of steps, as suggested by Leonard and Rayport [12]:

1. *Observation*. Observe individual customers or target user groups engaged in actual product use.
2. *Capturing data*. This can be done using a range of techniques as listed in Section 3.4.
3. *Reflection and analysis*. The data collected are discussed with the wider design team with the aim of identifying all the customers' requirements and converting these into design parameters (see Section 3.5).
4. *Generating solutions*. This is achieved using the common set of creative tools available to the designer, e.g. brainstorming, concept sketches, etc.

5. *Prototyping of chosen concepts.* Physical (or virtual) representations of the concepts that appear to meet most closely the customer needs are created.
6. *Verification of design.* Once again customers are brought in to evaluate the new designs to verify that they are an improvement on existing products.

This process will be iterative as several attempts may be needed before the new designs are seen to meet (and hopefully exceed) customer expectations.

The advent of RM means that customer expectations can be met more precisely through the design and production of customised products. The main benefit of customisation is higher value products that customers will pay more for. The correct balance must be reached between the number of features that are to be customised and the number of variations that are produced. This will be different for each product. In general, manufacturers should concentrate on customising the features that can add most value in comparison to the cost of customisation. The design and manufacturing of customised products introduces new issues to the operation of a business. If these are not dealt with in a coordinated manner, any price benefit delivered by customisation could be more than swallowed up by higher logistical, legal and product support costs. Therefore, customisation should be seen as part of a company-wide product strategy and not simply as a bolt-on extra to individual products.

## References

1. http://www.bbc.co.uk/insideout/east/series4/clive_sinclair_spectrum_c5.shtml (accessed July 2004).
2. http://www.retro-trader.com (accessed July 2004).
3. Langford, J. and McDonagh, D. (2002) *Focus Groups: Supporting Effective Product Development*, Taylor & Francis, London.
4. Akao, Y. (ed.) (1990) *Quality Function Deployment: Integrating Customer Requirements into Product Design*, Productivity Press, New York.
5. Evans, M.A. and Campbell, R.I. (2003) A comparative evaluation of industrial design models produced using rapid prototyping and workshop-based fabrication techniques, *Rapid Prototyping Journal*, **9**(5), 2003, 344–51.
6. http://wohlersassociates.com/3D-Digitizing-and-Reverse-Engineering.html (accessed July 2004).
7. Cain, R. (to be published) Involving users in designing: a framework based on understanding product representations, PhD Thesis, Loughborough University.

8. Porter, C.S. and Porter, J.M. (2000) Co-designing: designers and ergonomics, in *Collaborative Design* (eds S.A.R. Scrivener, L.J. Ball and A. Woodcock), Springer, London.
9. Khalid, H.M. (2001) Can customer needs express affective design?, in Proceedings of the International Conference on Affective Human Factors Design, Asean Academic Press, London.
10. Burns, A., Evans, S., Johansson, C. and Barrett, R. (2000) An investigation of customer delight during product evaluation, in Proceedings of the 7th International Product Development Management Conference, Leuven, Belgium, 29–30 May 2000.
11. Burns, A.D. and Evans, S. (2000) Insights into customer delight, in *Collaborative Design* (eds S.A.R. Scrivener, L.J. Ball and A. Woodcock), Springer, London.
12. Leonard, D. and Rayport, F.F. (1997) Spark innovation through empathic design, *Harvard Business Review*, November–December, 1997.

# 4

# CAD for Rapid Manufacturing

Rik Knoppers
*TNO Industrial Technology*
and
Richard Hague
*Loughborough University*

## 4.1 Introduction

The first commercial computer aided design (CAD) systems were sold in the 1970s and were initially driven forward by the needs and demands of the automotive and aircraft industries. The developments in the space and rocket sectors also contributed but to a lesser extent. Much of the initial effort was centred on the creation of two-dimensional drawings. These early CAD systems did not have any three-dimensional functionality and were developed to replace traditional transparencies, ink pens and drawing boards. Over time, CAD systems have evolved from two-dimensional into three-dimensional modellers through wire-frame, surfaces and solid-modelling systems, with solid modelling now becoming the norm.

As all additive manufacturing techniques are entirely dependent on the three-dimensional CAD description of the geometry to be built, this chapter will focus solely on the three-dimensional aspects of CAD systems. Some background will be given on the origin and the basics of CAD and then the relationship between CAD and Rapid Manufacturing (RM) will be clarified. Lastly, the future needs of CAD systems will be considered, which will give a view on the requirements of CAD to enable most to be made of the geometrical freedoms associated with RM.

*Rapid Manufacturing: An Industrial Revolution for the Digital Age*
Editors N. Hopkinson, R.J.M. Hague and P.M. Dickens 

## 4.2 CAD Background

### 4.2.1 History of CAD

Since computer memory was a rare and very expensive medium in the early days of CAD, the challenge was to find a mathematical description of the geometry that was as effective as possible. This is not a major concern as long as the geometry is assembled from primitive shapes (blocks, cylinders, spheres, cones, etc.). However, most products cannot be represented by such simple shapes and therefore, for ergonomic and aesthetic reasons, more complex three-dimensional curved shapes are desirable.

Since the automotive played an important role in CAD development, the generation of three-dimensional curved shapes has always been a driving force in the development of CAD systems. This resulted in the development of NURBS (non-uniform rational B-spline) which represented a very efficient mathematical description of a three-dimensional curved line and subsequent surface. However, some calculations with these mathematical descriptions are difficult to achieve or at least are not very efficient, and for this reason some operations (such as milling path generation and shading) were carried out by using other internal formats, such as 'tiles' or 'triangles' that are generated from NURBS. Therefore, a significant contribution in the performance of a CAD system is by the use of other internal formats other than NURBS. However, the NURB or formats resembling NURBS are still at the heart of today's CAD systems and thus the development of the NURBS is discussed in more detail below.

### 4.2.2 NURBS

To understand the basics of NURBS, we have to understand what the needs of the user are. The user – a designer of, for example, a new car – would like to have a smooth shape. To get a smooth shape, starting with, for example, a rectangular block, the first step is to round the corners. With the rectangular corners, the two surfaces of the shape are not tangential. After rounding these with a fixed radius, the surfaces are nicely tangential, but the radius is not continuous. The flat surface has an infinite radius while the rounded corner has a finite radius. This is demonstrated in Figure 4.1.

Therefore, despite the fact that the surface is nicely tangential, there is a discontinuity in the radius. Thus, though this has arguably improved the design, the blending will not look particularly good in reality as the discontinuity of the blend and the flat areas will give an annoying reflection

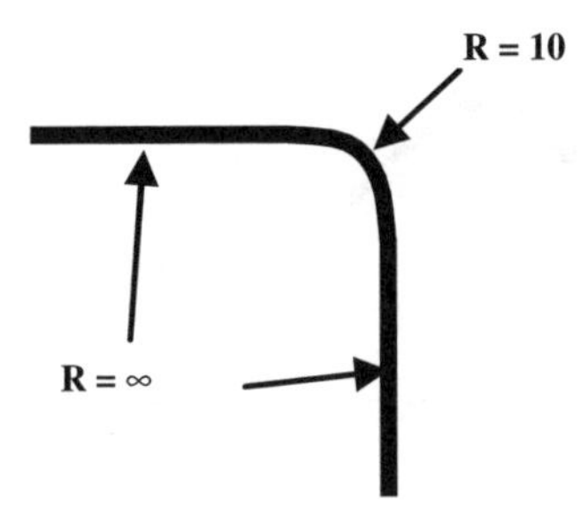


**Figure 4.1** Tangent rounding with discontinuous radii

of light on the surface. For this reason, work started on mathematical solutions to create shapes without a discontinuity and the solution was found by Pierre Bézier. Working at the Renault car manufacturer, from 1960 he started on the modelling of surfaces which resulted in the CAD system 'Unisurf'. In this CAD system the Bézier curve was used. The Bézier curve can be visualized as an elastic rubber connection between adjoining surfaces. The radius now smoothly varies from infinity at the flat surfaces to a minimum radius and then smoothly growing to infinity once more. This is demonstrated in Figure 4.2.

Picturing the Bézier curve as a rubber elastic band with a variable length, the shape of the curve can then be controlled by the endpoints of the curve and the point of intersection of the two lines tangential to the curve at the endpoints. These Bezier curves enabled designers to create very nice smooth curves and surfaces and were subsequently used in many CAD packages in the coming years.

However, as soon as a number of Bézier curves were needed to describe a shape, controlling the tangency between the Bézier curves became extremely difficult. This led to the development of B-spline curves, which can be thought of as a mathematical connection of a series of Bézier curves. This is demonstrated in Figure 4.3.

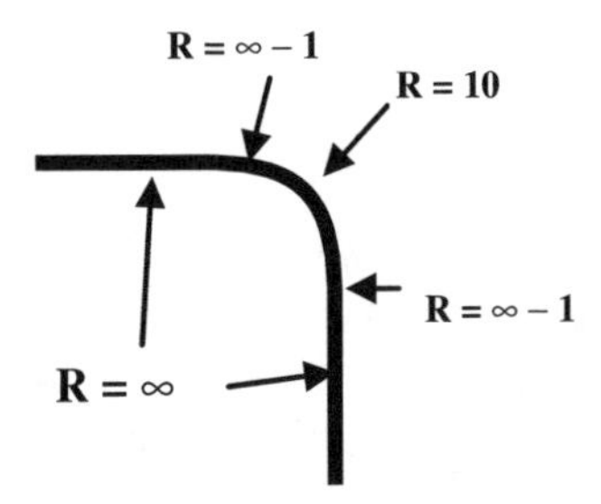


**Figure 4.2** Tangent rounding with continuous radii

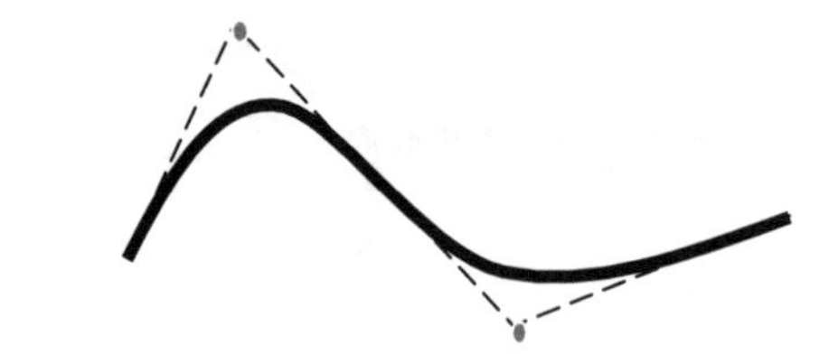

**Figure 4.3** B-spline curve with control points

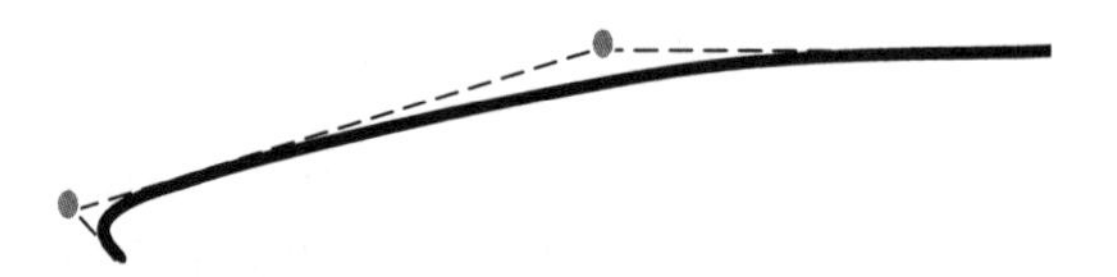

**Figure 4.4** Non-uniform B-spline

In order to design, for example, the bonnet of a car, a smooth soft rounding is also needed on 90% of the surface, and the rest should be curved more strongly. This means that in certain areas there should be more control points than at the smooth curved areas. For this, the non uniform B-spline was invented. This is shown in Figure 4.4.

In a CAD system, all the surfaces are described by the same mathematics, which results in not only complex curved surfaces being described by the method detailed above but also 'simple' geometries such as flat planes or even cylinders. However, approximating a cylinder with a non-uniform B-spline requires quite a considerable amount of control points. Therefore another optimisation of the B-spline was needed. This optimisation can be visualized by putting more or less pressure on the elastic band of each Bézier curve independently. This is shown in Figure 4.5. This property is called 'rational' and brings us to the non-uniform rational B-spline – more commonly known as 'NURBS'.

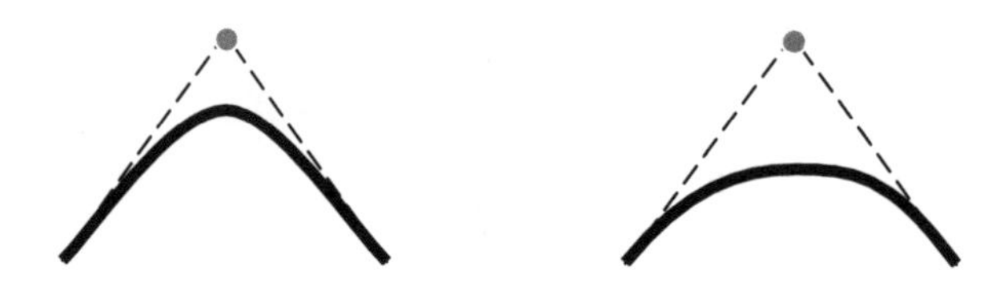

**Figure 4.5** Different rational effects on a Bézier curve

## 4.3 Relations between CAD and Rapid Manufacturing

### *4.3.1 From NURB to Rapid Prototyping and Rapid Manufacturing*

When stereolithography was introduced in 1988 the STL (Standard Triangulation Language) file was chosen as a neutral format between the CAD systems and the software supporting the stereolithography system. The format was easily accepted in the following years, since most CAD systems internally already used triangulation for various reasons. The STL format is now defined and accepted as a neutral format for all the RP (Rapid Prototyping), RT (Rapid Tooling) and RM systems.

The quality of the STL file is not always perfect due to the fact that it depends on the quality of the underlying NURB surfaces and hence the generation of the triangles on these surfaces. However, there are enough STL repair tool-kits available to repair these STL files and increase their quality. From these STL files, the RP, RT and RM systems create their own format to feed the machine. Some of them use slice contour information, some pixels per slice, some transfer their data to G–codes (a format used on milling machines) or any other slice information.

## 4.4 Future Developments Serving Rapid Manufacturing

Though CAD systems are already evolving in many aspects, such as ease of learning, more interconnection between design and engineering functions, lower price, etc., RM will require more drastic changes, which will strike at the essentials of today's CAD systems. The new opportunities afforded by RM, especially the design freedoms that are inherent with the additive approach to manufacture, will require a new approach to CAD systems in the coming years.

Three-dimensional CAD has developed with conventional manufacturing in mind and thus the complexity of the parts that can be conventionally made is somewhat limited. Consequently, the requirement for highly complex CAD designs is almost unnecessary. However, we are now at the point were the RM systems have the ability to produce virtually any complexity of geometry, which is counter to the ability of current three-dimensional CAD. Therefore, in order to maximise the potential of RM, a new CAD environment will need to be created. In particular, four different main directions for improvements are needed and include:

- Free feature modelling
- Product specific CAD
- Repeating features
- CAD for graded materials

### *4.4.1 Free Feature Modelling*

Three-dimensional CAD systems can be subdivided into two main principals: surface modelling and the so-called solid modelling systems. In surface modelling the user is able to define complex three-dimensional surfaces. By carefully generating and trimming these surfaces, it is possible, though not easy, to create 'solids' from them. To teach an individual to design and engineer products with a surface modeller, a considerable amount of time needs to be invested. In comparison, it is relatively easy to learn to create products with a good solid modeller. Modern solid modellers are easier to handle because they are dedicated to generating products using standard operations (bosses, holes, chamfers, fillets, drafts, etc.). However, this is also the reason why the geometrical freedom is more limited. Thus in conventional CAD, one could imagine working on a base block where, with simple imaginary mechanical actions like drilling, milling, chiselling, etc., material is removed (and also added!) as required.

However, in RM, the design of the part should escape the mechanical limitations and therefore the required CAD system should support other standard operations to create new types of features more suited to the needs of the designer of RM parts. Largely due to the fact that we have been restricted for years to 'Design for Manufacture', we do not yet have a full understanding of all the features that may be required for a 'typical' RM part – although we do know now that there is a need for complex three-dimensional new features. Such features could include three-dimensional honeycombs and assembly features including, for example, snap connections and screw treads. An example of an RM part that has had screw threads built in is shown in Figure 4.6; this approach will be common in the era of RM whereas it is difficult and costly to manufacture conventionally.

To optimise CAD for free feature modelling the basics of the CAD system can be preserved, but in the area of user interaction, features and standard operations substantial changes are needed.

### *4.4.2 Product Specific CAD*

RM challenges conventional wisdom in that historically the bottleneck in the manufacture of a part traditionally comes from the production of the mould tooling. In an age of RM, where no tooling exists and parts can be produced on demand in a matter of hours (or in the future, minutes), generation of the CAD model is actually the bottleneck. Therefore, the speed of the generation of the underlying three-dimensional CAD data will increasingly need to be addressed. One option for this would be in the creation of a product-specific CAD.

**Figure 4.6** RM part produced by the Perfactory™ process with three screw thread holes

There are already some examples of these kinds of development. Well-known examples include the number of specialised CAD systems for the design and engineering of the hearing aids detailed in Chapter 12, and the special 'CAD' facilities utilised by Invisalign for the creation of their custom dental aligners [1]. One can also find this in other sectors such as the footwear industry, where there are a number of developments to design footwear automatically on tailored CAD systems. This approach is also found in more technical domains such as the development of a semi-automatic tool design software by Materialise [2] and a fully automatic mould generation by CCIM (Competence Centre for Innovative Manufacturing) [3].

It is expected that this will be an exploding phenomenon in the coming years. Surprisingly, the well-established CAD vendors are not particularly involved in this work. Most of the work is executed by new emerging companies. Another remarkable thing is that most of these new applications are not built upon NURB-based systems but are using polygon-based systems with the STL format as an input.

### *4.4.3 Repeating Features*

As discussed previously, RM opens new unparalled design opportunities. One of these opportunities is the possibility of repeating features at a micro or macro level. Examples of repeating features are macro textures and textile replacements. Although the mapping of textures over surfaces is

**Figure 4.7** A potential product for a product dedicated CAD (rowing helmet Olympic Games 2004. Courtesy of DSM)

not new, these textures are in most cases an optical illusion and are not actual three-dimensional structures but are generated by lofting bitmaps on to the surfaces of CAD models. Conventional CAD is not capable of replicating such geometries. Examples of macro textures are shown in Figure 4.8.

To be able to Rapid Manufacture these structures, it is necessary that the structure itself is represented in a true three-dimensional geometrical form. Some of the problems of creating these types of mapped geometries were outlined in Chapter 2 where the problems of mapping textile replacements were discussed – this is similar to the

**Figure 4.8** Examples of macro textures

mapping of textures as they contain geometrical descriptions that are repeated in a certain pattern. However, the difference is that the macro textures become a part of the surface of the product; the textile replacements only use the surface as a guiding surface and are not connected to them.

However, this said, the problems in terms of CAD are similar and the main issue is simple to explain. Create a thousand or even a hundred complex small features and one will notice that conventional CAD systems will find it difficult or impossible to process, and this occurs with a number of problems. The data needed to mathematically describe the total geometry becomes extremely large and, additionally, the displaying (rendering) of such a complex total structure can also take a great deal of time.

Therefore, to enable the use of macro textures and replacement textiles to be possible within a CAD environment, the CAD has to be improved for the following aspects:

1. *Definition of the textures.* The definition of the texture geometry should be stored separately from the main product geometry. A relationship between the texture geometry and the product geometry should be defined. In this way the texture can be placed and oriented on the surface.
2. *Positioning and orientation of the texture.* Tools to create organised or random grids over three-dimensional curved surfaces should be available to the user. The user should also be able to define the orientation of the textures in a global way, e.g. all in one direction, perpendicular to the surface, all focusing at one point, etc.
3. *Reduction of the data volume.* Reduction of the data volume is required to enable the use of repeating features in three areas:
   - Storage of textured models
   - Graphical user interface to be able to display textured models in real-time
   - Communication speed with the RM machine

To enable repeating features on three-dimensional CAD systems it is likely that the fundamentals of today's three-dimensional CAD systems will have to be improved. This will involve a radical change in CAD which will not be easily acheived. The paradox in this will be that repeating features will be developed as soon as there is a big need for this kind of software, while these repeating features can only be created efficiently using this software. Figure 4.9 shows and example of a repeating structure that has been produced via selective laser sintering (SLS).

**Figure 4.9** Example of an experimental macro texture produced by SLS

## 4.5 CAD for Functionally Graded Materials (FGMs)

Many of today's RM technologies create products in a point-by-point fashion. Selective laser sintering and stereolithography, for example, initiate the solidification at the point where the laser hits the surface. Also, some of the layer manufacturing technologies even deposit the material point by point. Examples of these technologies include the 3D Systems InVision and the Objet machines; both of these systems build a product drop by drop using a modified inkjet system. Both systems, when supplied with one or more extra jets (jetting other materials), could build graded material parts comparable with printing a stack of colour pictures on top of each other. The creation of such products is know as 'functionally graded materials' (FGMs) and this type of graded internal structures is only possible when taking an additive approach to manufacturing.

However, one of the main issues for FGMs is that it is impossible to represent these in a CAD environment. Most commercial CAD systems are in the category of 'B-rep' modellers. B-rep stands for 'boundary representation', which implies that these systems describe the geometries only by the definition of their outer surface, not only for surface modellers but also for the so-called solid modellers. The inside of the 'solids' on these CAD systems are as empty as the volume surrounding the solid. Therefore, for most CAD systems, even though they are known as 'solid modellers', they are, in fact, just surface representations of the geometry. This is also true of

**Figure 4.10** Cut through a CAD solid model showing the 'empty' inside

the STL file format. An example of what the internals of a conventional CAD 'solid' model look like is shown in Figure 4.10.

Therefore, with commercial CAD systems, we have no method at present to represent the internal structure/composition at individual points within the volume of the part. Of course, this has not been necessary before as we have had no way of producing FGMs, but now that we do with RM, a new method of data representation is required. The future 'FGM machine' needs to be provided with information to 'print' the desired graded material distribution. When investigating alternatives that are available to describe internal composition, a number of options are available. These include:

- Voxels
- Finite element model (FEM) elements
- Particle system elements
- Vague discrete modelling elements

However, all these types of entities have the major disadvantage of using massive amounts of memory and also some of them are rather too complex in their execution or do not have any relation between the entities, which in turn adversely influences the possibilities of editing these kinds of format. The following section will discuss voxels in more detail.

### *4.5.1 Voxel-Based FGMs*

FGMs could simply be solved by sending the machine (for each layer) a sheet of pixels – similar to printing colour pictures. These stacked layers of pixels are known as voxels (volumetric pixels). An example of voxel representation of clouds is shown in Figure 4.11.

Voxels originated from the medical field as they are used in MRI and CT scans. These medical voxelated data sets are only generally required for display purposes – no manipulation of them is required and they are

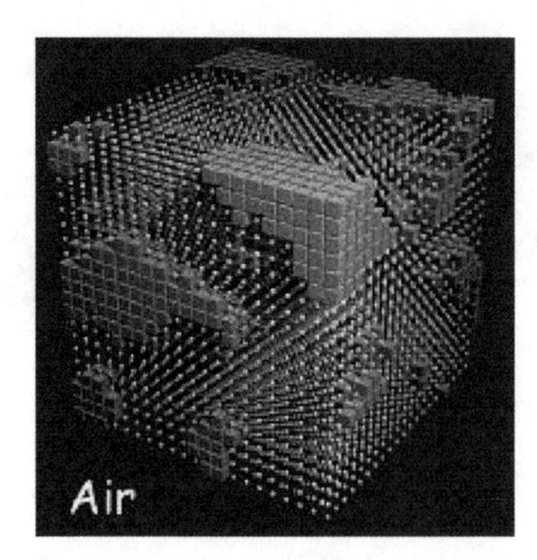


**Figure 4.11** Voxel representation of clouds

created directly from scan data. However, as in an FGM CAD environment, we will need to build in the functional grading, this then gives rise to the problem of how voxel-based information can be produced in a CAD environment.

Though voxels are very practical to use as neutral data sets between the design system and the machine fabricating the part, they are not the ideal entity to design and edit graded information. The methodology of how to create and edit functionally graded designs on a CAD system and how this new type of volumetric property design (VPD) system, i.e. both non-NURB and non-STL based, could perform its functions are discussed below.

### *4.5.2 VPD System*

The best way of understanding what a VPD system should be able to do is to analyse the situation from the user's point of view. Imagine a producer of sweets developing a new kind of candy stick. The instructions are:

1. The candy stick should grade from being sour at one end of the stick to sweet at the other end.
2. The candy stick should also grade from being red at the outside to white at the inside.

This is shown diagrammatically in Figure 4.12.

With these two instructions the model properties on each position have been described within the candy stick. The VPD system should now translate these desired model properties to material properties. The system must be able to give the value of any property at any point (sour, sweet, red, white) so that the stick can be produced, realising the correct material composition. This is shown in Figure 4.13.

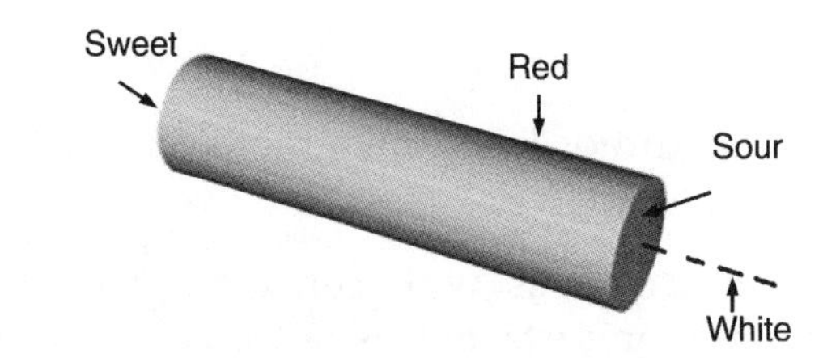


**Figure 4.12** Candy stick with desired properties

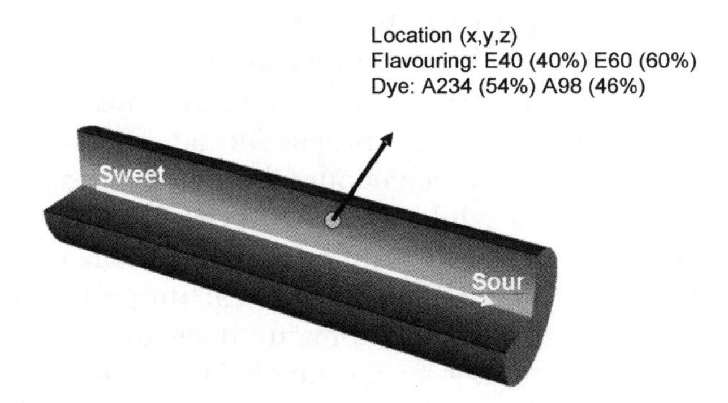


**Figure 4.13** Example of local properties in a candy stick

From this example we learn that it is impossible for the user to describe the material properties at any point. Properties are described in some specific areas where the properties are equal. These areas will be named 'ISO-props'. Using these ISO-props, the area between these properties is described by the transition function from one property to the other – this is known as the 'distribution function'. When calculating the distribution of materials it will be clear that this should be limited by the boundary of the solid. This means that we would like to limit the distribution of a property by defining a domain. Therefore, all together, volumetric property distribution of materials can be defined by:

- Domains
- Properties
- Distribution functions

## Domains

Ultimate freedom of volumetric property distribution can be created with multiple domains. Every distribution function is valid in a specific domain. The properties used by the distribution function are not necessarily located in the domain.

## Properties

Intensive research by the author has shown that it is necessary to define two types of properties:

1. *ISO-props.* ISO-props are areas with constant properties. They can be defined by a point, line or surface. It is not possible to intersect or touch two ISO-props as this would lead to an internal conflict. For example, a point cannot be at the same time defined as black and white. The example of the candy stick has four ISO-props of two types: taste (sweet, sour) and colour (red, white). Properties of the same type are able to influence each other while properties of a different type are independent of each other.
2. *VARI-props.* For more advanced property distributions, ISO-props are not sufficient. For this we need VARI-props. Although a VARI-prop and an ISO-prop do have the same behaviour, the ISO-prop describes an area with constant properties while the VARI-prop describes an area with variable properties. The VARI-prop is dependent on other ISO-props. When on a point of the VARI-prop the local properties are needed to radiate these properties in the domain, then this local property is calculated from the distances to the involved ISO-props. The involved ISO-props are not distributing their properties to the domain, exclusively to the VARI-prop.

## Distributions

A distribution function describes the quantity (e.g. percentage) of a property as a function of the distance. To be able to define effective distributions, two distribution functions are defined:

1. *Absolute distribution function.* The absolute distribution function is defined by a function and the distance to a prop. The function gives the relation between a relative distance and a percentage of the property, the distance is a given value by the user. An example of an absolute distribution function is given in Figure 4.14.

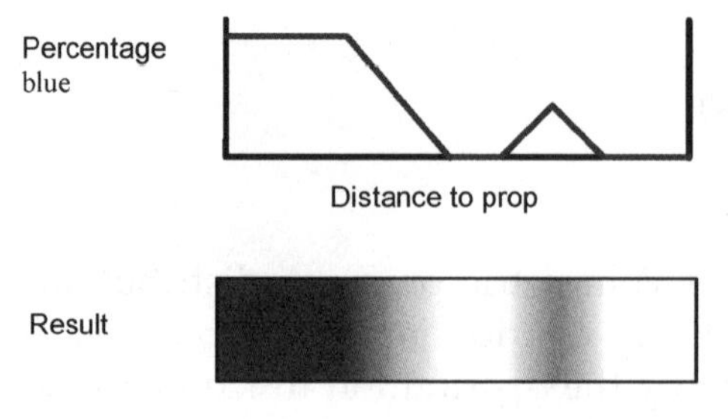


**Figure 4.14** Example of an absolute distribution function

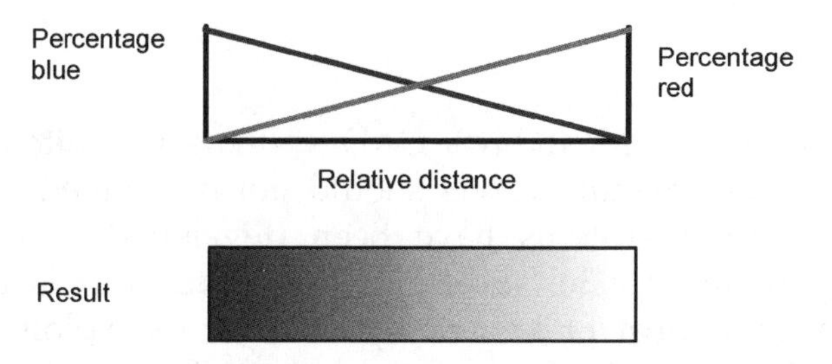


**Figure 4.15** Example of a relative distribution function

2. *Relative distribution function.* The relative distribution function is defined by two props and a function. From a given point the shortest distance is calculated to both props. From this the relative position of the point is calculated. The relative position is used in the function to calculate the values of the properties. An example of a relative distribution function is given in Figure 4.15.

### *4.5.3 Summary of FGMs*

This proposed set-up for a CAD system to design and edit graded materials is certainly not the only solution to solve this problem. However, by evaluating this approach, it can be seen that it is vital to create a new concept in CAD for graded materials if we are able to produce complex internal compositions like those shown in Figure 4.16.

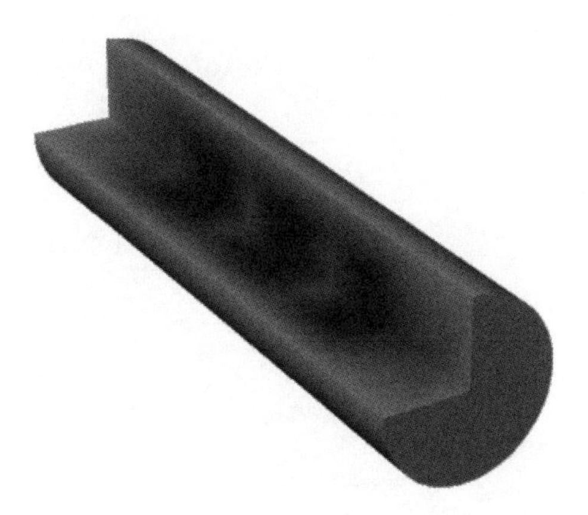

**Figure 4.16** Example of a complex graded material structure

## 4.6 Conclusion

Rapid Manufacturing with today's CAD systems is limited but possible, though CAD system manufacturers should not be blamed for this limited functionality as their systems have been developed with conventional manufacturing in mind. The subject of design for RM is new and thus there is not a great deal of knowledge of how to exploit fully the vast potential that RM offers in the area of design. The development of CAD systems for RM and the knowledge for design and engineering of RM products are dependent upon each other and are expected to mutually stimulate improvements and developments.

Parallel to this, new CAD oriented knowledge-based systems will be developed to design customer/product dedicated items efficiently. In this way, the costs of design for RM will be effectively managed, thus making RM products more affordable. This tendency towards product dedicated CAD systems will strongly stimulate the developments in RM.

'Repeating' features such as real textures are not really possible with today's CAD. To fully exploit the possibility of repeating features, the core technology of today's CAD systems needs to change. This will not easily be done.

Additionally, new CAD systems are needed for functionally graded materials. It is expected that this development will take place separately from the other developments in CAD for Rapid Manufacturing.

## References

1. www.invisalign.com
2. www.materialise.com
3. www.ccim.nl

# 5

# Emerging Rapid Manufacturing Processes

Neil Hopkinson and Phill Dickens
*Loughborough University*

## 5.1 Introduction

To those who have worked in the field of Rapid Prototyping, Tooling and Manufacture for many years, a number of processes, such as stereolithography and selective laser sintering, may be considered as established rather than emerging. However, with respect to the field of Rapid Manufacture these processes are very much in their infancy primarily because the field of Rapid Manufacture is itself immature. Consequently, this chapter will discuss processes that were originally intended for Rapid Prototyping/ Tooling and are currently being used for Rapid Manufacture as well as those that have been and are being developed with Rapid Manufacture in mind.

In order to illustrate the infancy of current Rapid Manufacturing (RM) technologies it is worth making a comparison between selective laser sintering (SLS) of polymers and one of today's foremost polymer manufacturing technologies – injection moulding. Figure 5.1 shows some of the key milestones in the development of injection moulding and selective laser sintering of polymers. The injection moulding development timeline shows a vast gap between the first widely recognised patent granted to the Hyatt brothers in 1872 [1] and widespread adoption of the technology. The slow uptake in technology was largely due to a lack of suitable materials, with

*Rapid Manufacturing: An Industrial Revolution for the Digital Age*
Editors N. Hopkinson, R.J.M. Hague and P.M. Dickens 

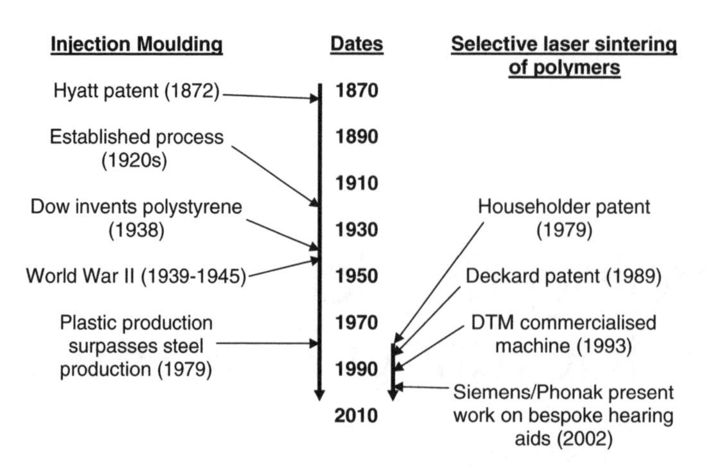


**Figure 5.1** Key milestones in the development of injection moulding and selective laser sintering

highly flammable cellulose-based materials being the first in use. The invention of new materials such as polystyrene created a push for the technology and this, coupled with an unprecedented pull from requirements for high volumes of products during World War II, resulted in an upsurge of applications for injection moulding. By 1979 plastic production had surpassed the staple of the industrial revolution, steel, with further developments continuing to date [2].

The selective laser sintering of polymers development timeline in Figure 5.1 is clearly shorter than that for injection moulding, highlighting the infancy of the technology. However, it should be noted that a mere 23 years after Ross Householder's patent, the applications for selective laser sintering were far wider than those for injection moulding in 1895 (23 years after the Hyatt brothers' patent).

At present there are over 20 different recognised Rapid Prototyping (RP) technologies [3], but not all of these can be considered suitable for Rapid Manufacture as some have material properties that render them useless for anything beyond visualisation. This is a subjective issue and may be contentious in some cases, but this chapter will only consider processes that create parts that are suitable for applications beyond visualisation. This is a grey area as certain materials may or may not be deemed suitable for applications beyond visualisation; for example, parts made by the Z-Corp process that have been infiltrated with epoxy may have some limited functional capability. Another example may be in the manufacture of parts that are only intended for visualisation in end use; for example, a sculpture created in wax by the ThermoJet process would have very limited mechanical properties but may be entirely adequate for its intended use.

The technologies discussed in this chapter have been divided into three different categories according to the raw material used in the process. These categories are:

- Liquid-based systems
- Powder-based systems
- Solid-based systems

Within each category the technologies are presented in terms of their maturity, so commercialised systems will precede those that are being developed in laboratories. The list of technologies covered in this chapter is not exhaustive but represents those that either currently show the best promise for widespread use to manufacture end-use products or those that have the best potential to become widely used for RM in the future.

For those technologies that have not yet been commercialised, the nature of processing has been taken into consideration. In order to achieve high throughput and lower cost for end-use manufacture, and especially series manufacture, RM systems of the future will be ones that process material within a layer simultaneously (using '1D' arrays or '2D' matrices) rather than sequentially (using '0D' spots). In order to explain this nature of processing, Figures 5.2 to 5.4 depict simplified versions of '0D', '1D' and '2D' methods for processing material.

Figure 5.2 shows a plan view of how stereolithography uses a single spot laser to cure resin. The spot is considered to be '0D' (in reality it has an area of $\pi r^2$ – typically 0.03 mm$^2$) and needs to scan in *X* and *Y* in order to cover the area of each layer. Figure 5.3 shows how multi-jet modelling uses a '1D' (one-dimensional) array of ink jets to simultaneously deposit material, e.g. to bind together particles in the three-dimensional printing process. In this case, as the array is in one dimension (*Y*) the print heads only need to traverse in one axis (*X*). Figure 5.4 shows how the digital mirror device (DMD) used in the Perfactory™ process allows a two-dimensional matrix of

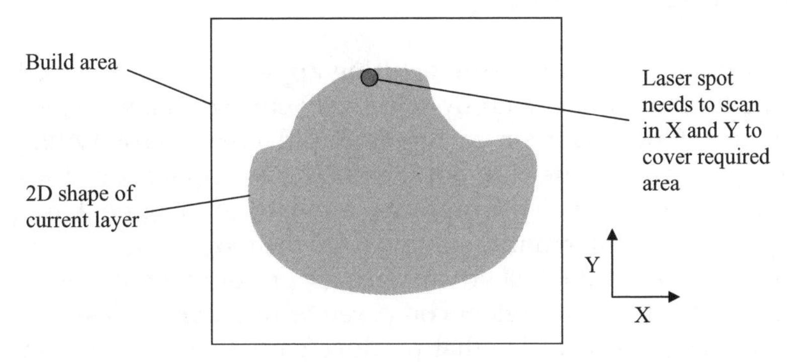


**Figure 5.2** '0D' processing by a laser in the stereolithography process

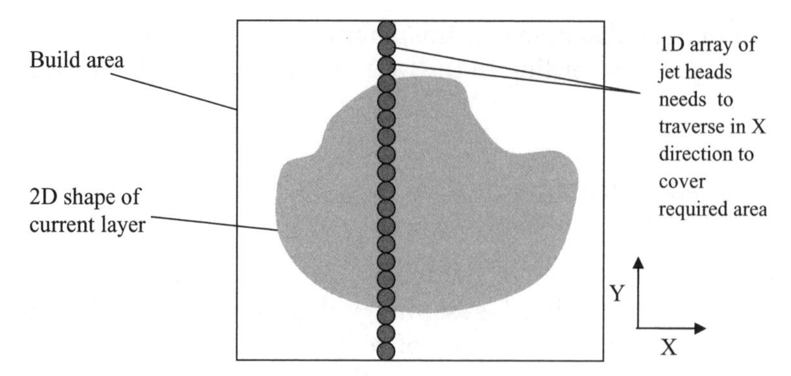


**Figure 5.3** '1D' processing by multi-jetting in the Z-Corp three-dimensional printing process

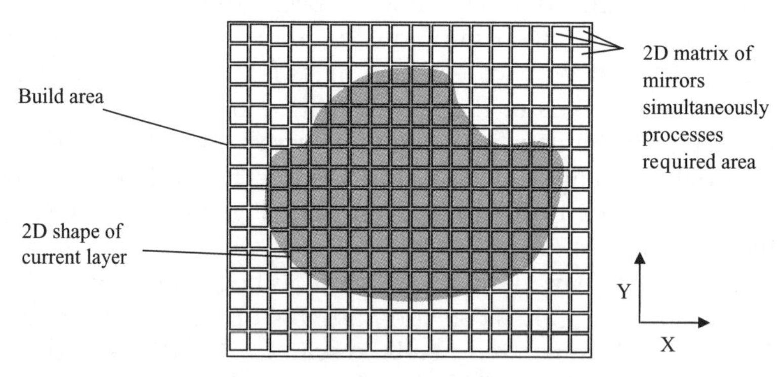


**Figure 5.4** '2D' processing by multiple mirrors in the Perfactory™ process

selected mirrors to reflect light on to the resin surface. In this case there is no requirement for scanning or traversing.

## 5.2 Liquid-Based Processes

Liquid-based layer additive manufacturing approaches, almost exclusively, involve the formation of a solid by selectively curing regions of photosensitive polymers. One exception to this is Rapid Freeze Prototyping, which creates ice parts from water [4], but the likely RM applications for ice parts beyond visualisation (e.g. ice sculptures) are hard to imagine. Liquid-based technologies, notably stereolithography, were the pioneering processes in RP and have a number of distinct advantages for prototyping, including superior accuracy and definition when compared with other processes. In recent years a number of filled resins that produce parts that look very much like injection moulded components have been commercialised. The fact that these components have a similar appearance to injection moulded parts may help to ease the decision to adopt RM with these materials. However, the material properties of photocured parts tend to be relatively poor when compared with other processes, especially over a period of time when ageing, for example, by exposure to sunlight, which causes continued curing, can severely affect mechanical properties and appearance. Sensitivity to humidity can also be a problem with photocurable resins. Photocurable resins that can achieve stable properties over time and in different environments would provide a significant new set of potential applications for RM.

### 5.2.1 Stereolithography

This is widely considered as the founding process within the field of RP, with the first patent granted to Chuck Hull in 1986 leading to the first commercial machine from 3D Systems in 1987 [5]. Figure 5.5 shows a schematic of how the stereolithography process works using an ultraviolet (UV) laser to initiate a curing reaction in a photocurable resin. Using a computer aided design (CAD) file to drive the laser, a selected portion of the surface of a vat of resin is cured and solidified on to a platform. The platform is then lowered, typically by 100 μm, and a fresh layer of liquid resin is deposited over the previous layer. The laser then scans a new layer that bonds to the previous layer. In some areas where overhangs are created,

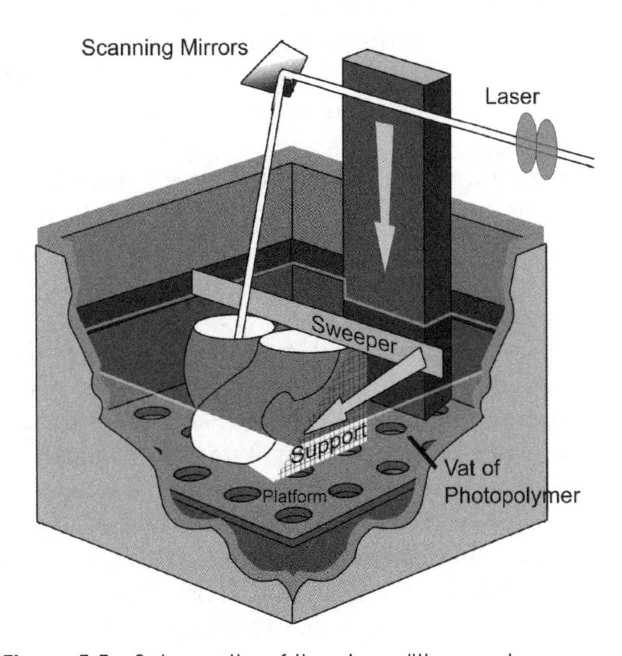


**Figure 5.5** Schematic of the stereolithography process

supports are automatically generated by the machine's software. These supports may be edited by the operator prior to building and need to be removed once the final part is made. After building, parts are removed from the machine and platform, supports removed and post-processing in a UV and/or thermal oven are used to cure any uncured resin.

### *5.2.2 Jetting Systems*

Two different technologies that use inkjet technology to create parts from photocurable resins have been commercialised. These are the PolyJet™ process from Objet of Israel and the InVision process from 3D Systems.

The PolyJet™ process developed by Objet of Israel was first 'announced' to the RP community in the late 1990s and was commercialised in April 2000 with the Objet Quadra machine. The process uses an array of printing heads to simultaneously selectively deposit an acrylate-based photopolymer. Each layer, which can be as thin as 16 μm, is then cured by a trailing UV lamp that passes over the deposited material. Supporting material is simultaneously jetted through a second series of jets and cured to a gel state with the UV lamp, so that it may be removed by water jet or similar after building. As with other processes at the time, the process was initially aimed at the RP/T (Tooling) market, but the emergence of RM applications has resulted in an interest in using the technology for end-use products. Haim Levi from Objet suggests that the RP market place is becoming saturated and has stated that 'Rapid manufacturing will be the breakthrough for this technology'. Levi considers the technology as suitable for RM in terms of accuracy, resolution and speed, but acknowledges that material properties are the current weakness [6].

The InVision machine was commercialised by 3D Systems in 2003 and was originally intended as a means of producing RP parts in the timescales offered by the ThermoJet process, which produces wax parts, and with material properties approaching those from stereolithography. The process shares many features with the Objet process, including the use of an array of jets to print an acrylate-based material (first used in stereolithography in the 1980s). One obvious difference between InVision and Objet's PolyJet™ process is the support structure – with InVision the supports are created by jetting wax which can be removed by a variety of means after parts are built. Coloured resins allowing parts to be made in a single colour (e.g. red, blue, black, etc.) have proved to be popular and should help with respect to achieving improved aesthetics from parts. As with the Objet process, the main obstacle for widespread use in Rapid Manufacture is the material properties of the parts made. Figure 5.6 shows parts made by the InVision process.

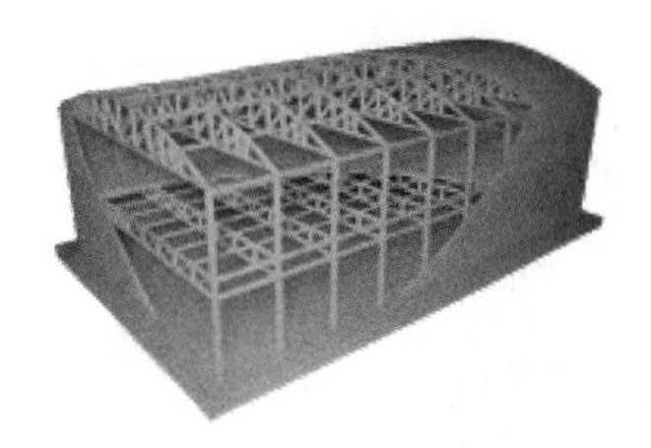
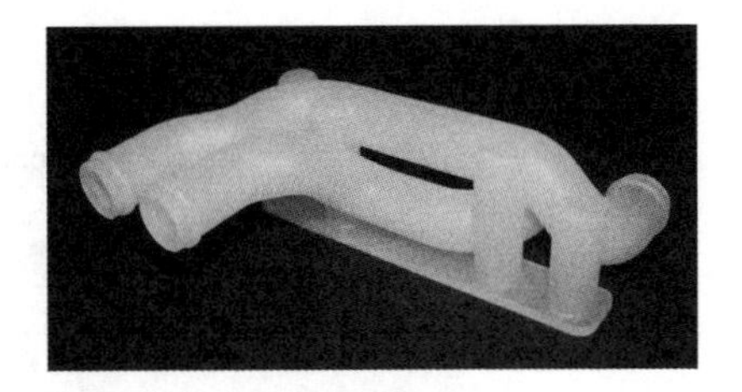

**Figure 5.6** Parts made by the InVision process. (Reproduced with permission of 3D Systems, Inc.)

### 5.2.3 Direct Light Processing™ Technologies

Digital mirror devices (DMDs) developed by Texas Instruments have found a wide variety of applications ranging from data projectors to the manufacture of electronic products. In terms of layer manufacturing technologies as covered by this text, the Perfactory™ process developed by EnvisionTec of Germany is the only machine available commercially. The Perfactory™ machine was first commercialised by EnvisionTec in March 2003 and is a particularly interesting technology from the perspective of Rapid Manufacture. Interestingly, the name of the machine (an abbreviation of 'personal factory') implies that it is intended to make products (factory) and that these products are likely to be customised to the individual (personal). Beyond the name, the process also has a number of distinguishing aspects, starting with the fact that the process builds parts that 'grow' downwards rather than upwards (see Figure 5.7).

As with PolyJet™ and InVision the process builds parts from an acrylate-based photocurable resin, but it does so by using a two-dimensional matrix of mirrors rather than a '1D' array of print heads to selectively cure the material. In order to selectively cure a layer, the process makes use of DMD technology developed by National Instruments to selectively switch on and off mirrors that reflect UV light from a source on to the build area. Figure 5.8 shows a number of parts made by the Perfactory™ process. With a build speed of 10–15 seconds per layer the process is well suited to building parts quickly, but the use of a single DMD with a finite matrix of pixels limits the process to small parts if a fine resolution is maintained. Given the suitability to produce small parts, it is of little surprise that the hearing aid industry has shown significant interest in this technology with a number of machines supplied to US-based manufacturers.

### 5.2.4 High-Viscosity Jetting

The principle involves continuous change in a layer's pattern (negative image of the layer) according to a very thin slice of the object to be printed.

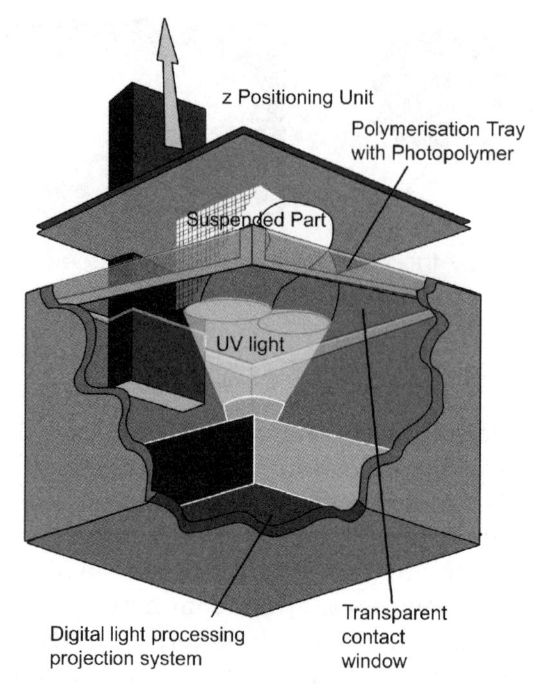


**Figure 5.7** Schematic of the Perfactory™ process

**Figure 5.8** Parts made by the Perfactory™ process

This uses a mechanism based on displacing a small drop of a printable material (powder-filled polymer paste) to a desired location on a substrate. The fundamental unit consists of a single jet, which is controlled by air jet pressure, the distance from the substrate and the length of the jetting pulse (see Figure 5.9). An experimental programme on single jets is being carried out and the results are showing the different shapes and sizes of deposition that can be achieved. This concept will be scaled to a block of multi-jets controlled in parallel to deposit a layer of a desired pattern. The final process will provide solutions to a number of problems and limitations known in conventional printing and existing RP machines. It also has flexibility in the

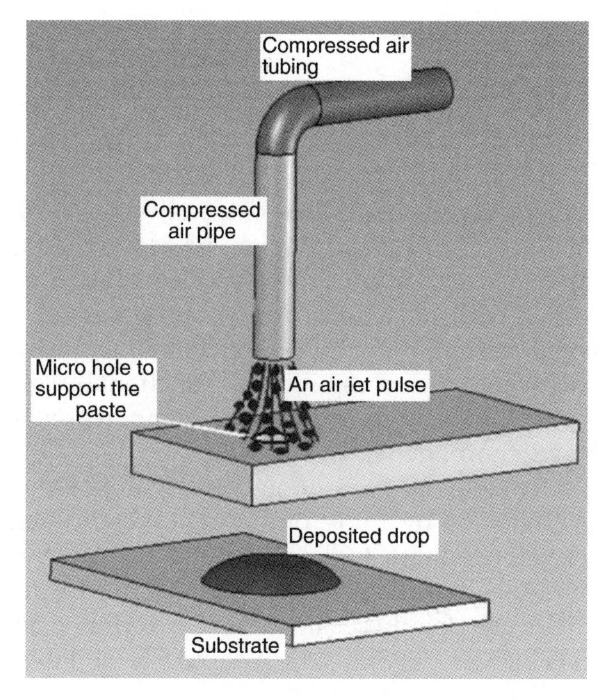


**Figure 5.9** The high-viscosity jetting principle

degree of accuracy depending on the hole size being used for the jet. A production speed similar to existing high-volume production methods will be possible and the paste can be loaded with any powder.

### *5.2.5 The MAPLE Process*

MAPLE DW (matrix assisted pulsed laser evaporation: direct write) was invented by researchers at the Naval Research Laboratory, Washington. It uses a high-repetition-rate, 355 nm UV laser beam which is focused on a transparent material or 'ribbon' that has a 1–10 μm thick layer of build material on the underside (see Figure 5.10). As the laser energy is directed to the ribbon the build material transfers to the receiving substrate. This is analogous to a typewriter ribbon.

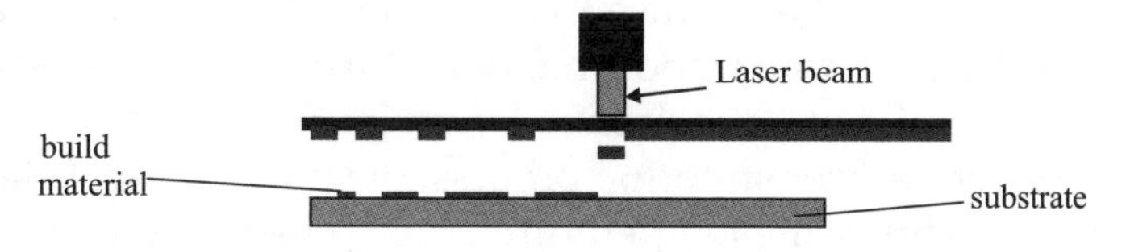


**Figure 5.10** Matrix assisted pulsed laser evaporation

## 5.3 Powder-Based Processes

The material properties and stability of parts that may be achieved with powder-based processes means that they will, in the long run, be more suited to RM than the liquid-based systems (this is being borne out in current RP systems where powder-based selective laser sintering is more widely used for Rapid Manufacture than its closest liquid counterpart, stereolithography). Powder-based processes also offer a wider variety of material possibilities with polymers, metals and ceramics all available on current commercial systems. Furthermore, combing powders and layer additive manufacturing, the possibility of functionally graded materials (see Chapter 7) provides a unique potential for increased functionality of rapid manufactured components.

The technologies covered in this section begin with polymer selective laser sintering which is followed by a host of different metal-based processes that were arguably developed with tooling in mind but are beginning to find applications in end-use manufacture. Finally, a series of polymer powder-based processes that use one-dimensional and two-dimensional processing, as described earlier in this chapter, which should be suitable for medium to high volume manufacture, is described.

### *5.3.1 Selective Laser Sintering (Polymers)*

Selective laser sintering was first invented and patented by Ross Householder in 1979, but it was only commercialised following the work of Carl Deckard at the University of Texas at Austin in the late 1980s. This led to the formation of the DTM Corporation who first commercialised a machine in 1992 and developed the technology to allow the processing of a variety of polymers along with ceramics and metals (see the next section). In 1994 EOS GmbH released their EOSINT machine, which still enjoys a significant market share [5].

The process is in many ways similar to stereolithography, but the powdered raw material is sintered or melted by a laser that selectively scans the surface of a powder bed to create a two-dimensional solid shape. A fresh layer of powder, typically 100 μm thick, is then added to the top of the bed so that a subsequent two-dimensional profile can be traced by the laser bonding it to the layer below. The process continues to create a full three-dimensional object and the un-fused powder acts as a supporting material which obviates the need for support removal during post-processing. Figure 5.11 shows a schematic of the selective laser sintering process.

During the selective laser sintering process, the powder bed is heated prior to laser scanning to bring the temperature of the powder up to a temperature that is typically a few degrees Centigrade below the sintering temperature.

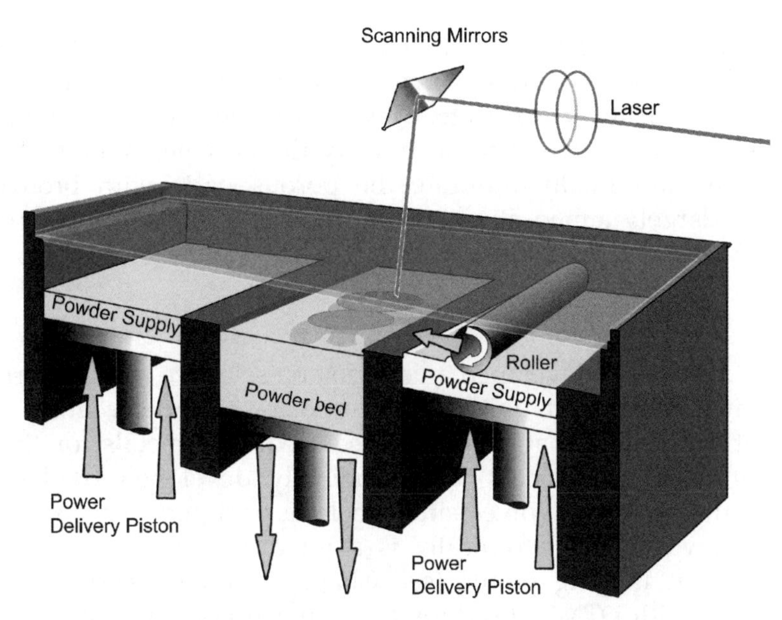


**Figure 5.11** Schematic of the selective laser sintering process

This pre-heating is usually performed by infrared heaters and helps the process by reducing thermal gradients between sintered and non-sintered powder and reduces the energy required by the laser to sinter the powder. Highly crystalline polymers – notably nylons – are sintered by using the laser to elevate the powder temperature to the melt temperature ($T_m$). This leads to good contact between particles and hence parts with relatively good mechanical properties. These good mechanical properties have allowed semi-crystalline polymers to be used in numerous RM applications to date. Amorphous materials, such as polycarbonate, do not have sharp melting points and are sintered by using the laser to elevate the powder temperature to the glass transition temperature ($T_g$). This leads to weaker parts than those that are sintered at the melt temperature but such parts have been widely used as investment casting patterns that do not require high strength. The low strength of amorphous laser sintered parts is likely to restrict their potential uses in RM applications.

### *5.3.2 Selective Laser Sintering (Ceramics and Metals)*

During the 1990s both DTM and EOS developed the selective laser sintering process by allowing complex cores and moulds for sand casting applications to be created using sand particles coated with a polymer (resin) binder. Various work has also been done on selective laser sintering of ceramic parts, but this has not yet been fully commercialised.

DTM also applied the concept of using coated powders to metals so that the selective laser sintering machine could produce powder metallurgy steel parts in the green state. These parts could then be subjected to post-processing in a furnace to burn away the polymer binder, sinter the steel particles and finally infiltrate the porous parts with bronze. This process was largely aimed at producing tooling but offers some potential for Rapid Manufacture of end-use products.

### *5.3.3 Direct Metal Laser Sintering*

During the 1990s EOS developed a variation on selective laser sintering that could produce metal parts without the need for a binder coating and the subsequent processing that would be required. The metals for the direct metal laser sintering process were originally developed by Electrolux. Essentially the process involves either melting or liquid phase sintering of the metal powder, which typically is a mixture of various components having different melting points (see Chapter 7 on Functionally Graded Materials). As with DTM's development of metal laser sintering, the initial goal of direct metal laser sintering was to produce tooling, but the process has been used for end-use Rapid Manufacture (the term used by EOS for this is 'e-Manufacturing$^{TM}$').

### *5.3.4 Three-Dimensional Printing*

The three-dimensional printing process (see Figure 5.12) invented at MIT was licensed to ExtrudeHone to allow the development and sale of metal parts predominantly for tooling applications. Using '1D' jetting technology

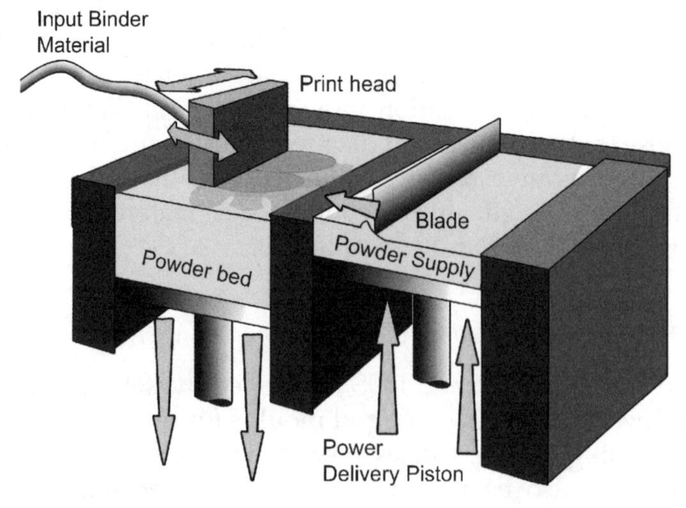


**Figure 5.12** Schematic of the three-dimensional printing process

the process has a relatively high throughput in terms of creating green parts similar to those by metal selective laser sintering described above. Post-processing is similar to that for selective laser sintered parts, but surface finish usually requires some form of machining to create a surface suitable for tooling. In terms of RM, there have been no published examples but the process may be suited to more rigorous applications where polymers will not suffice, especially where fine surface finish is not required. MIT has also granted licenses to Therics for medical applications and to the Specific Surface Corporation, who have used three-dimensional printing to create ceramic filters.

### *5.3.5 Fused Metal Deposition Systems*

A number of processes have been developed that use the principle of blowing metal powders into a melt pool created by a laser (see Figure 5.13). Among the developers of these technologies were Sandia National Labs who used the expression laser engineered net shaping (LENS) [7] and joint work between John Hopkins University, Penn State University and the MTS Systems Corporation [3]. Different versions of the technology have been commercialised by numerous organisations including POM, Optomec and

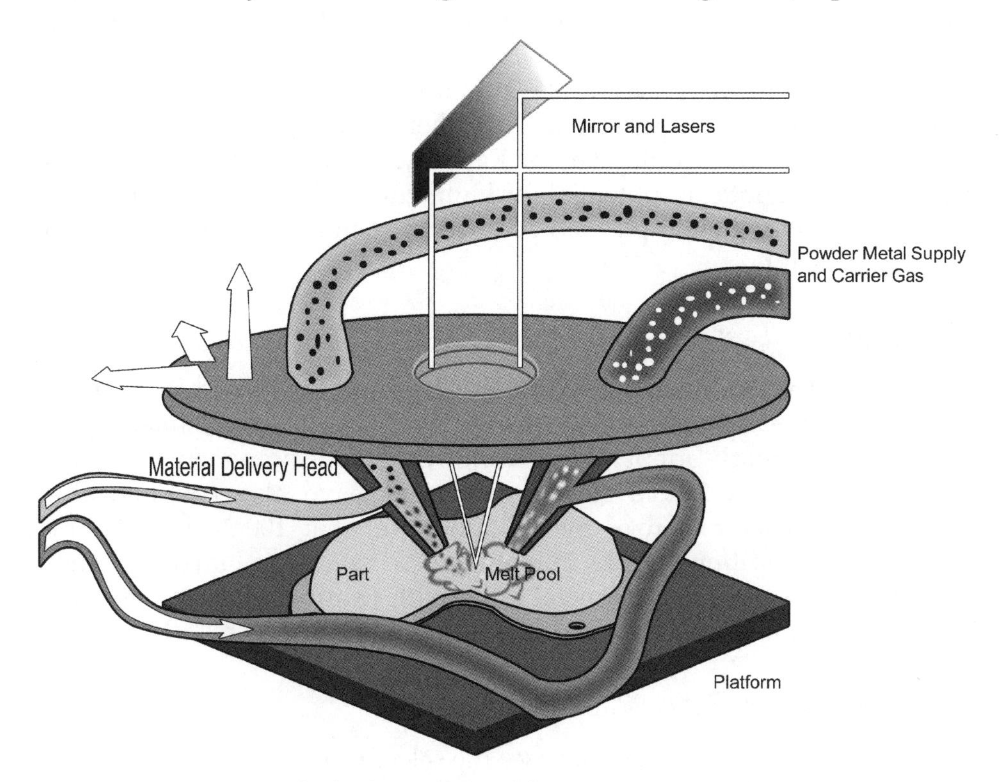


**Figure 5.13** Schematic of fused metal deposition

Aeromet. Generally these processes have relatively slow deposition rates and produce parts with poor surface finish, but they do offer the potential to process functionally graded materials (see Chapter 7) in high melt temperature metals including titanium. These processes have also proved to be particularly adept at fixing broken parts such as mould tools by adding material where required. This may form an RM niche for these processes in the comparably high-added-value area of product repair/maintenance.

### *5.3.6 Electron Beam Melting*

The electron beam melting (EBM) process was first commercialised by Arcam in Gothenburg, Sweden, in 1997 [8]. The process uses a similar approach to selective laser sintering but replaces a laser with an electron beam – this has interesting implications. Firstly, the electron beam may be directed by changing the electromagnetic field through which it passes. This eliminates the need for scanning mirrors and can significantly increase scanning speed (up to $1\,\text{km}\,\text{s}^{-1}$). Secondly, the power developed by the electron beam is very high, allowing the process to fully melt a wide range of metals including titanium alloy using a very fast scanning rate. However, the process is limited to conductive materials and surfaces, as with many other layer-based processes, often require extensive finishing – especially for tooling applications.

Although the process uses a '0D' scanning approach, the speed of scanning coupled with no requirement for further furnace processing may make the process a leading contender for Rapid Manufacture. In particular, the process offers significant potential for high temperature or medical applications, especially with complicated geometries such as the knee joint shown in Figure 5.14 [9].

### *5.3.7 Selective Laser Melting*

MCP Group have commercialised the Realizer machine that uses a laser to fully melt stainless steel parts in a similar manner to laser sintering. The process is particularly adept at producing very small components, including ones with complex lattice structures. Trumf have also commercialised machines for laser fusion of metals and claim that 100 % densities may be achieved. Additionally 3D Micromac AG have developed the microsintering process to produce metal parts with feature resolution as fine as 20 μm using submicrometre-sized powder that is sintered by a laser [10].

### *5.3.8 Selective Masking Sintering*

The Swedish company Speedpart uses the selective masking sintering (SMS) technology that has been developed by Ralf and Ove Larson in Sweden. The

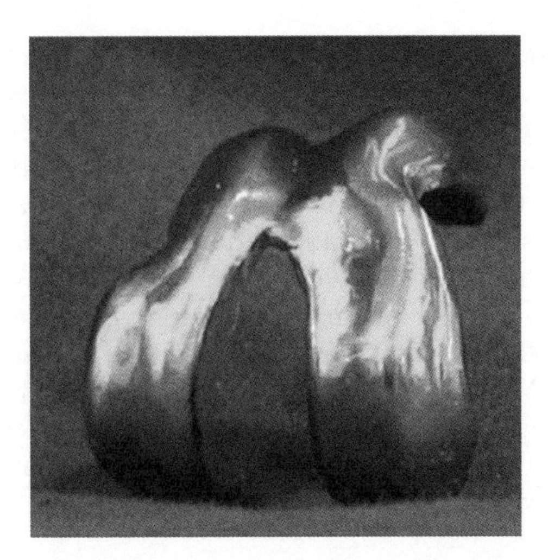

**Figure 5.14** EBM model of a knee joint made from H13 tool steel. (Reproduced by permission of NC State University from O. Harrysson *et al.*, Direct fabrication of metal orthopedic implants using electron beam melting technology, in Proceedings of the 14th Solid Freeform Fabrication (SFF) Symposium, Austin, Texas, 4–6 August 2003, pp. 439–46)

SMS process involves printing a mask of infrared radiation reflecting material on to a glass sheet and placing the sheet over a powder bed. Infrared radiation is then applied to the glass sheet and allowed to selectively pass through the mask in order to sinter the powder directly below. A schematic of the SMS process is shown in Figure 5.15.

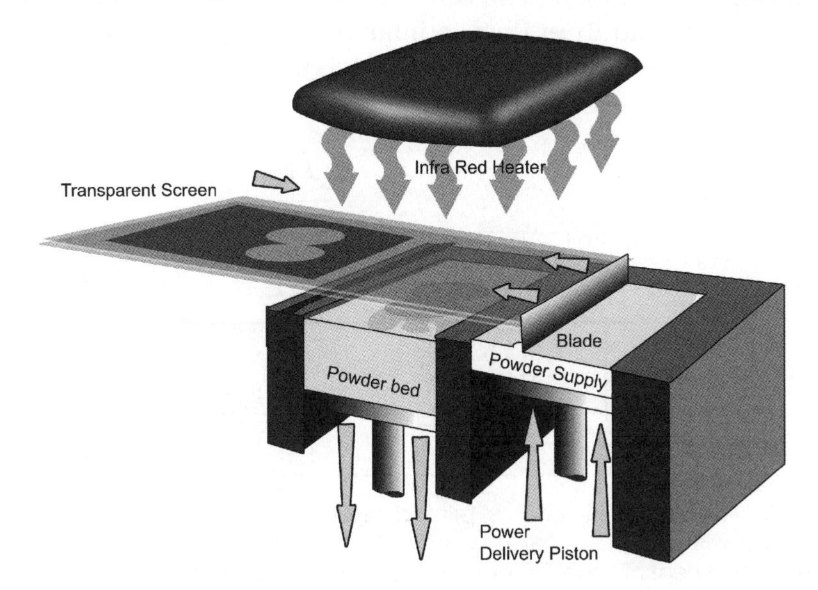


**Figure 5.15** Schematic of the SMS process

This process eliminates the requirement for a laser and in instances where a significant portion of the surface needs to be sintered this should dramatically reduce processing times when compared with selective laser sintering. Speedpart claim that each layer can be fully processed in 10–20 seconds and that the use of a mask in place of a laser ensures that build times are easy to predict and independent of part volume. Consequently, this approach should have maximum benefits when being used for Rapid Manufacture in high volumes. To date one machine, the RP3 with a build volume of 300 mm × 210 mm × 500 mm, can be used to produce parts from one material, VT3 a glass-filled nylon powder. The process was initially aimed at producing vacuum forming tools, making use of the process's inherent porosity, but new materials may make this one of the next generation of RM machines.

### *5.3.9 Selective Inhibition Sintering*

The selective inhibition sintering (SIS) process is being developed by Behrokh Khoshnevis at the University of Southern California with Rapid Manufacture in mind. Rather like selective masking sintering from Speedpart, the process seeks to combine the benefits of SLS (material properties) and jetting processes (build speed) to address two of the major concerns behind Rapid Manufacture. However, Khoshnevis suggests that SIS is likely to achieve better resolution and definition than the Speedpart process as the inhibiting material is printed directly on to the powder and uses no mask that might allow for light diffusion [11]. The process was described at the Solid Freeform Fabrication Symposium in Austin, Texas in August 2002 [12] and uses a print head to jet fluid to inhibit sintering on to selected areas of the build volume (see Figure 5.16). This is followed by using a radiating heat

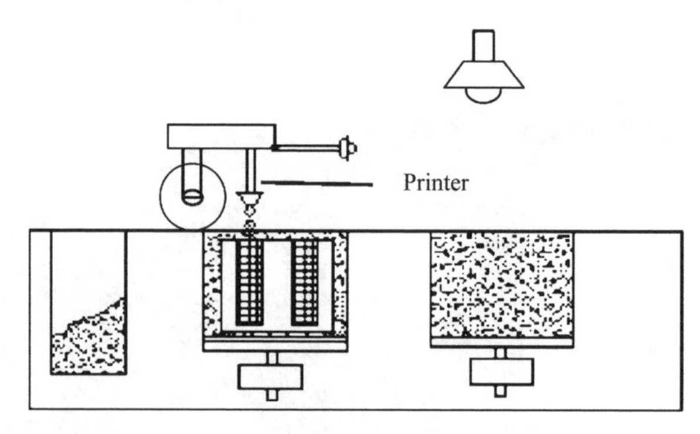


**Figure 5.16** A sintering inhibitor is printed selectively on to the powder surface. (Reproduced with permission of University of Southern California from asiabanpour *et al.*, Advancements in the SIS process, in Proceedings of the 14th Solid Freeform Fabrication (SFF) Symposium, Austin, Texas, 4–6 August 2003, pp. 25–38)

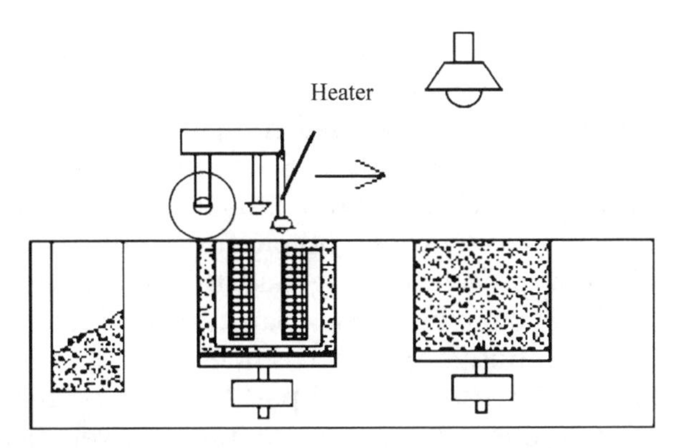


**Figure 5.17** Radiant heat source sinters powder that has not had the inhibitor printed on to it. (Reproduced with permission of University of Southern California from asiabanpour *et al.*, Advancements in the SIS process, in Proceedings of the 14th Solid Freeform Fabrication (SFF) Symposium, Austin, Texas, 4–6 August 2003, pp. 25–38)

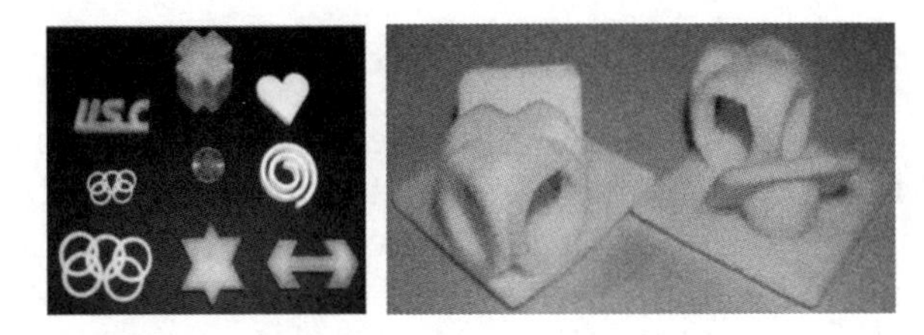

**Figure 5.18** Parts made by the SIS process. (Reproduced with permission of University of Southern California from asiabanpour *et al.*, Advancements in the SIS process, in Proceedings of the 14th Solid Freeform Fabrication (SFF) Symposium, Austin, Texas, 4–6 August 2003, pp. 25–38)

source to traverse the build area and sinter any powder that has not had the inhibitor printed on to it (see Figure 5.17).

Initial work used a single nozzle to print around the edge of parts but the process could easily be developed to simultaneously print the inhibiting material in a '1D' array or possibly a two-dimensional matrix. Figure 5.18 shows parts made by the SIS process.

Recent research has considered the use of a variety of inhibiting materials ranging from commercial cleaning agents to potassium iodide. Another aspect of materials that has been researched is the powder material of the parts themselves, with success reported when sintering a variety of polymer powders including polystyrene, polycarbonate and polyester [13].

Unlike most of the other powder sintering processes, SIS does not require that the material comprising the part be elevated to a higher temperature than the material not to be sintered – this may result in a reduction of thermal gradients across the surface which may lead to benefits such as reduced warpage. The process is in many respects a mirror of other processes such as three-dimensional printing in that the material that is printed on to the surface is used in areas that will not constitute the final part. In most cases the actual volume of a part that is processed compared with the volume that it occupies is quite small. Consequently, SIS will need to apply the inhibiting material to the majority of each layer with parts only comprising a minority of each layer. From the perspective of high-volume manufacture this appears to be a counterintuitive approach. Having said this, the goal of high-volume manufacture by RM will often be to pack part beds as densely as possible, so this apparently counterintuitive approach is likely to be less of an issue than it would be for RP, where densely packed part beds are seldom used.

Newer implementations of SIS incorporate power waste reduction mechanisms that act in the form of shutters that block selected areas of a passing heater bar from emitting heat to the powder surface underneath (see Figure 5.19). In this way, areas of powder treated by inhibitor as well as sintering of non-part powder sections will be minimized.

### *5.3.10 Electrophotographic Layered Manufacturing*

Ashok Kumar has been developing the electrophotographic layered manufacturing (ELM) process at the University of Florida [14]. This process uses an interesting mix of ideas that have been used for laser sintering. Figure 5.20 shows how the process uses electrophotographic methods to deposit a part powder and then a support powder for each layer.

Initial work focused around the idea of producing a green part by depositing separate part and support powders and then using a furnace operation to sinter the part material in a separate step; this required that the

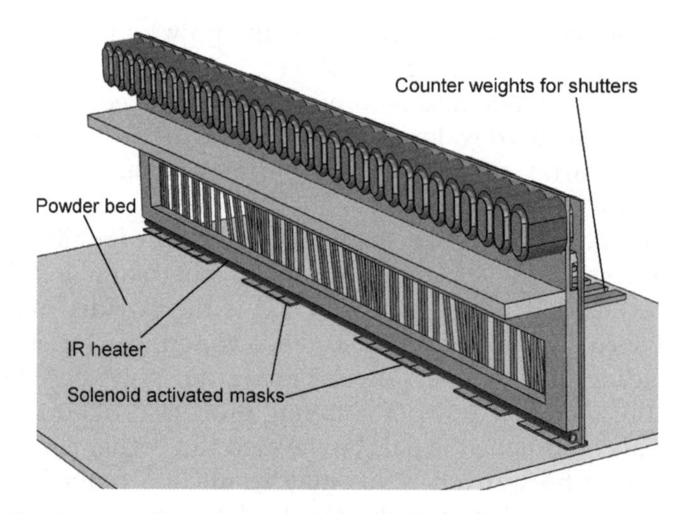


**Figure 5.19** The heater assembly showing some masked and unmasked areas under the heater. Shutters are solenoid activated. (Reproduced with permission of Behrokh Khoshnevis)

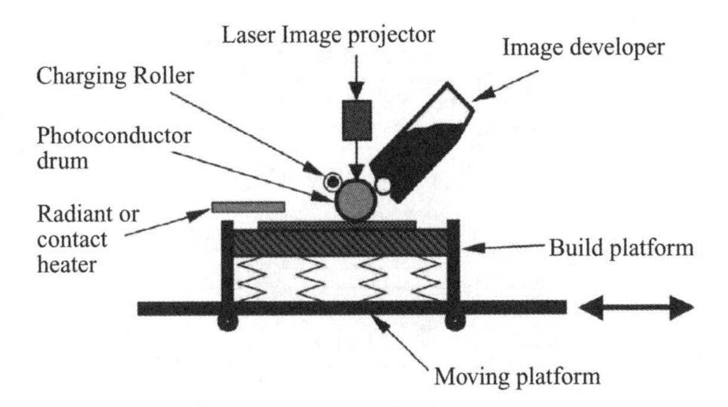


**Figure 5.20** The electrophotographic layer manufacturing process. (Reproduced with permission of *Rapid Prototyping Journal* from A.V. Kumar and A. Dutta, Investigation of an electrophotography based rapid prototyping technology, *Rapid Prototyping Journal*, 2003, **9**(2), 95–103, MCB University Press Limited)

**Figure 5.21** Parts sintered in the build bed made by electrophotographic layer manufacturing. (Reproduced with permission of *Rapid Prototyping Journal* from A.V. Kumar *et al.*, Electrophotographic printing of part and binder powders, *Rapid Prototyping Journal*, 2004, **10**(1), 7–13, Emerald Group Publishing Limited)

support material had a higher melt point than the part material. However, further work has experimented with the idea of sintering each layer before the next layer is deposited, as with other powder-based layer manufacturing processes [15]. Figure 5.21 shows some parts that have been sintered in the part bed. One of the problems that needs to be overcome is in depositing material electrophotographically to create parts with a large $Z$ height. It appears that the process could be suited to very high production rates but limited to smaller parts such as electrical components.

### *5.3.11 High-Speed Sintering*

The high-speed sintering (HSS) process is being developed at Loughborough University. As with the processes above, HSS is aimed at taking advantage

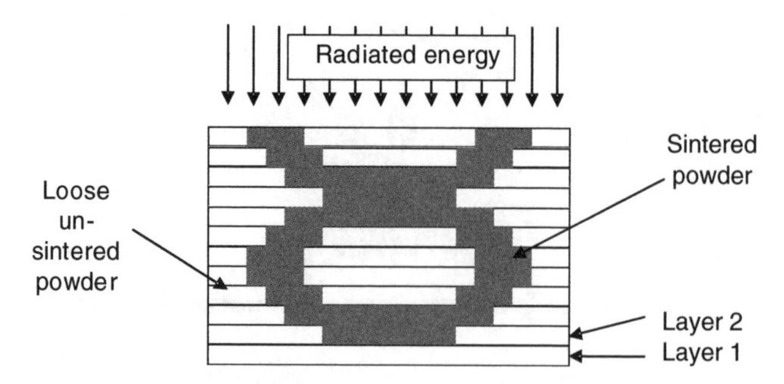


**Figure 5.22** Adding material to promote sintering to the powder bed to selectively sinter powder in the high-speed sintering process

of the mechanical properties given by SLS while achieving an increase machine throughput and reduced machine cost by eliminating the need for a laser [16]. HSS defines the geometry of each layer by printing a material that promotes absorption of radiation (and hence promotes sintering) on to the powder bed surface, rather like a negative of SIS (see Figure 5.22).

The key to HSS is the ability to control the rate of sintering across the build surface. Figure 5.23 shows how different amounts of intensity radiation and the addition of varying amounts of carbon black to nylon powder can alter sintering times [16].

Research has shown that a high sintering rate results in minimal shrinkage and good edge definition but poor mechanical properties, while slow

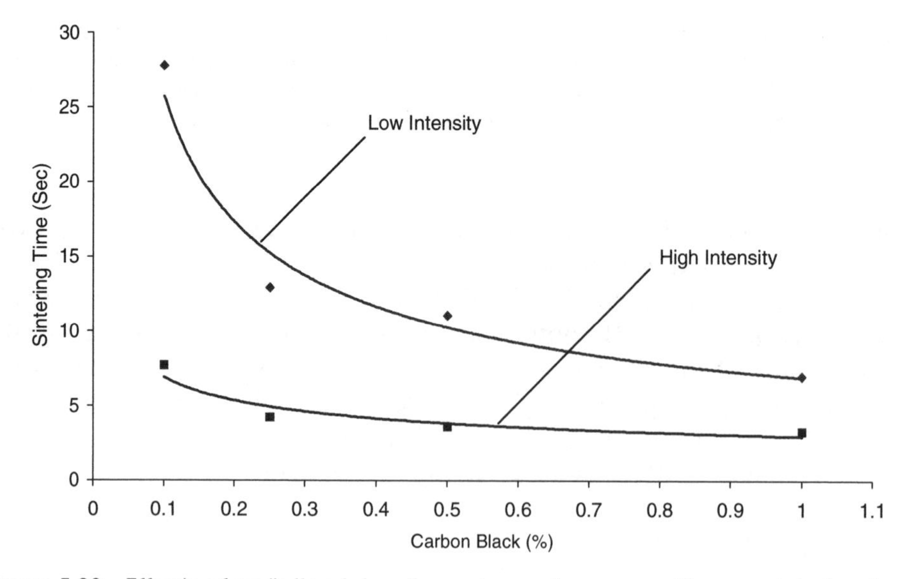


**Figure 5.23** Effects of radiation intensity and powder composition on sintering time

**Table 5.1** Mechanical properties of nylon-12 sintered by HSS and SLS

| Process | Composition of carbon black (% by weight) | Young's modulus (MPa) | UTS[b] (MPa) | Elongation at break (%) |
|---|---|---|---|---|
| SLS[a] | 0 | 1600 | 44 | 9 |
| HSS | 0.25 | 1633 | 47.5 | 18 |
| HSS | 2 | 1666 | 46.4 | 15 |

[a]Details obtained from www.3Dsystems.com.
[b]UTS, ultimate tensile strength.

sintering achieves better mechanical properties but at a cost of definition and accuracy. By controlling sintering rates via techniques such as the use of greyscale and materials that absorb energy at different rates the goal of achieving good mechanical properties with good accuracy and surface finish is being pursued. Table 5.1 shows the mechanical properties of parts produced by HSS and show that these exceed the properties from SLS. This suggests the suitability of the process for relatively demanding applications.

## 5.4 Solid-Based Processes

Processes that use a solid raw material in non-powder form have been an integral part of the RP industry since its formative years in the early 1990s. The two predominant forms of solid-based processes that are discussed here (fused deposition modelling and laminate object manufacturing) have been commercialised for some time, but incremental improvements continue by both the suppliers and academic institutions worldwide.

### *5.4.1 Fused Deposition Modelling*

This process was first commercialised by Stratasys in 1991 with patents awarded to Scott Crump, the company founder in 1992 [3]. Stratasys now have more machines installed than any other RP supplier (over 3000 globally) and in 2003 surpassed all previous annual unit sales, shipping 691 machines [5], although it should be stressed that many of these machines are at the cheaper end of the market than many other processes. The fused deposition modelling (FDM) process creates parts by extruding material (normally a thermoplastic polymer) through a nozzle that traverses in $X$ and $Y$ to create each two-dimensional layer. In each layer separate nozzles extrude and deposit material that forms the parts and material that form supports where required. The use of a nozzle with a diameter of typically $\sim$0.3 mm limits resolution and accuracy. Also the need for the nozzles to

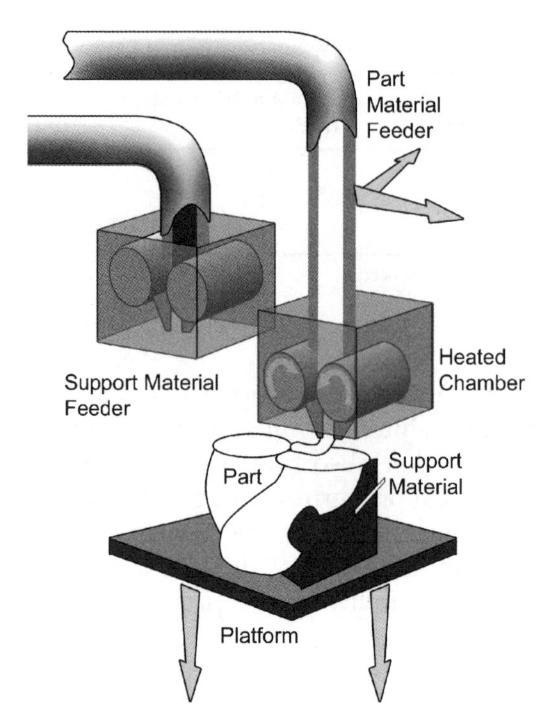


**Figure 5.24** Schematic of the fused deposition modelling process

physically traverse the build area limits build speed, but the process is very easy to set up and can operate in an office or factory environment. Support removal can be manual or, when water soluble supports are employed, they may simply be dissolved, the latter approach being most valuable with more complicated geometries. Figure 5.24 shows a schematic of the FDM process that can produce parts in materials including polycarbonate, polyphenylsulfone and, most commonly acrylonitrate butadiene styrene (ABS). The simplicity of the process should make it suitable for the development of a wide variety of thermoplastic polymers, which may open up opportunities for Rapid Manufacture. To date a few examples of Rapid Manufacture from FDM exist including gun mounts for the military [5] and pill dispensing hardware for the pharmaceutical industry.

An interesting development based on FDM is contour crafting (CC), a process invented at the University of Southern California by Behrokh Khoshnevis. CC exploits the surface-forming capability of trowelling in order to create smooth and accurate planar and freeform surfaces out of extruded materials – mainly construction ceramics. As shown in Figure 5.25, the extrusion nozzle has a side trowel. As the material is extruded, the traversal of the trowels creates smooth outer and top surfaces on the layer. The side trowel can be deflected to create non-orthogonal surfaces. The

**Figure 5.25** The contour crafting process. (Reproduced with permission of Behrokh Khoshnevis)

extrusion process builds only the outside edges (rims) of each layer of the object. After complete extrusion of each closed section of a given layer, if needed, filler material such as concrete can be poured to fill the area defined by the extruded rims. Figure 5.26 shows a large CC machine used to create objects with construction in mind.

**Figure 5.26** Large-scale contour crafting machine for construction. (Reproduced with permission of Behrokh Khoshnevis)

### *5.4.2 Sheet Stacking Technologies*

A number of technologies have been developed to create three-dimensional parts by cutting and stacking two-dimensional sheets of various materials. Different approaches have been used to cut sheets, bond them together and remove waste material from each sheet. Helysis first commercialised the laminate object manufacturing (LOM) process in 1991 with a number of similar organisations including KIRA, Kinergy, Solidica and Solidimension, following up with similar processes during the 1990s.

The Helysis process (now supplied by Cubital) involves stacking layers of paper with a bonding material and creating the part profile by cutting each layer of paper with a laser (see Figure 5.27). Post-processing involves using hand tools to remove the unwanted material and to reveal the part inside. The main problem for the process is that for complicated geometries, and

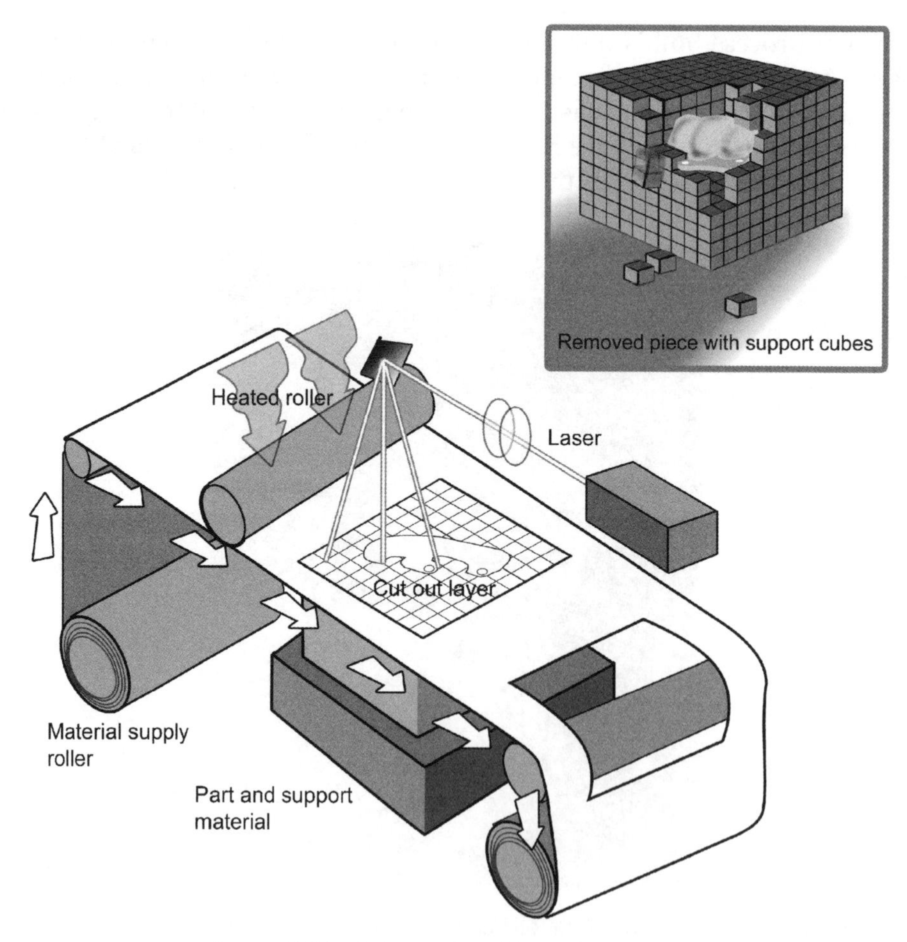


**Figure 5.27** Schematic of the LOM process

especially those with thin walls, post-processing is difficult, time consuming and can damage the part. For simple geometries this is less of a problem, but simple geometries are usually more suitably produced by machining. Variations on the process involve the use of a knife to cut the paper or the use of different materials.

Different materials provide the potential for sheet stacking technologies to be used in Rapid Manufacture. Solidimension use polyvinyl chloride (PVC) materials in their process, but more environmentally friendly polymers may be required to increase interest in Rapid Manufacture from their machines. Solidica use metal sheets that are bonded by low-temperature ultrasonic diffusion, with machining employed to cut out the required geometry of each layer. This process has some interesting value added potentials, such as the ability to embed fragile fibre-optic cables within parts [17].

## Acknowledgement

The authors are indebted to Joe Shaffery of ShafferyPayne, Global Creative Networks, for the creation of many of the images in this chapter.

## References

1. Rubin, I.I. (1973) *Injection Moulding Theory and Practice*, Society of Plastics Engineers, John Wiley & Sons, Ltd. Chichester.
2. Bryce, D.M. (1997) *Plastic Injection Molding*, Society of Manufacturing Engineers, Dearborn, Michigan.
3. Chua, C.K., Leong, K.F. and Lim, C.S. (2003) *Rapid Prototyping, Principles and Applications*, World Scientific Publishing Co. Pte. Ltd, Singapore.
4. Zhang, W., Leu, M.C. and Ji, Z. (1998) Rapid freeze prototyping with water, in Proceedings from the Solid Freeform Fabrication (SFF) Syposium, Austin, Texas, 10–12 August 1998, pp. 185–92.
5. Wohlers, T. (2004) Rapid prototyping and tooling state of the industry, Annual Worldwide Progress Report, Wohlers Associates Inc.
6. IEE Manufacturing Engineer (2004) *Rapid Prototyping: 'The Breakthrough Will Come'*, April–May 2004, IEE Publishing, UK, p. 7.
7. Griffith, M.L., Schlienger, M.E., Harwell, L.D., Oliver, M.S., Baldwin, M.D., Ensz, M.T., Smugeresky, J.E., Essien, M., Brooks, J. and Robino, C.V. (1998) Thermal Behaviour in the LENS Process, in Proceedings from the Solid Freeform Fabrication (SFF) Syposium, Austin, Texas, 10–12 August 1998, pp. 89–96.
8. Larsson, M., Lindhe, U. and Harrysson, O. (2003) Rapid manufacturing with electron beam melting (EBM) – a manufacturing revolution?, in

Proceedings of the 14th Solid Freeform Fabrication (SFF) Symposium, Austin, Texas, 4–6 August 2003, pp. 433–8.

9. Harrysson, O., Cormier, D.R., Marcellin-Little, D.J. and Jajal, K.R. (2003) Direct fabrication of metal orthopedic implants using electron beam melting technology, in Proceedings of the 14th Solid Freeform Fabrication (SFF) Symposium, Austin, Texas, 4–6 August 2003, pp. 439–46.
10. Regenfuß, P., Hartwig, L., Klotser, S., Ebert, R., Brabant, T., Petsch, T. and Exner, H. (2004) Industrial freeform generation of microtools by laser micro sintering, in Proceedings of the 15th Solid Freeform Fabrication (SFF) Symposium, Austin, Texas, 2–4 August 2004, pp. 710–19.
11. Wohlers, T. (2003). Rapid prototyping and tooling state of the industry, Annual Worldwide Progress Report, Wohlers Associates Inc.
12. Khoshnevis, B., Asiabanpour, B., Mojdeh, M., Koraishy, B., Palmer, K. and Deng, Z. (2002) SIS–a new SFF method based on powder sintering, in Proceedings of the 13th Solid Freeform Fabrication (SFF) Syposium, Austin, Texas, August 2002, pp. 440–7.
13. Asiabanpour, B., Khoshnevis, B., Palmer, K. and Mojdeh, M. (2003) Advancements in the SIS process, in Proceedings from the 14th Solid Freeform Fabrication (SFF) Symposium, Austin, Texas, 4–6 August 2003, pp. 25–38.
14. Kumar, A.V. and Dutta, A. (2003) Investigation of an electrophotography based rapid prototyping technology, *Rapid Prototyping Journal*, **9**(2), 95–103, MCB University Press Limited.
15. Kumar, A.V., Dutta, A. and Fay, J.E. (2004) Electrophotographic printing of part and binder powders, *Rapid Prototyping Journal*, **10**(1), 7–13, Emerald Group Publishing Limited.
16. Hopkinson, N. and Erasenthiren, P.E. (2004) High speed sintering – early research into a new rapid manufacturing process, in Proceedings from the 15th Solid Freeform Fabrication (SFF) Symposium, Austin, Texas, 2–4 August 2004, pp. 312–20.
17. Kong, C.Y., Soar, R.C. and Dickens, P.M. (2004) Ultrasonic consolidation technique for embedding SMA fibres within aluminium matrices, *Composite Structures*, **66**(1–5), 421–7, Elsevier Ltd.

# 6

# Materials Issues in Rapid Manufacturing

David L. Bourell
*The University of Texas at Austin*

## 6.1 Role of Materials in Rapid Manufacturing

Materials flexibility in Rapid Manufacturing (RM), along with accuracy and surface finish, has been a critical factor in the technology from the very beginning. It is an enabling feature of RM. As is the case for any manufacturing process, the choice of materials is in part dependent on the specifics of the process. Just as fluidity is critical for casting and plasticity is requisite for forging, RM processes impose constraints on the range of available materials. Coupled to process constraints is product demands desired by the final user of the product. A major role of materials developers in RM is the tailoring of materials, alloys and multi-component systems to create processable materials with an acceptable end performance.

## 6.2 Viscous Flow

The motion of liquid is important in many RM processes. Polymers melt and flow in fused deposition modelling (FDM) and selective laser sintering (SLS). Metals are molten in the powder spray processes and in direct laser sintering. Post-process infiltration of porous RM preforms also involves liquid flow. For all these, a critical feature of materials flow is the viscosity. The viscosity $\eta$ is for Newtonian flow, considered to be the constant of

*Rapid Manufacturing: An Industrial Revolution for the Digital Age*
Editors N. Hopkinson, R.J.M. Hague and P.M. Dickens 

proportionality relating the shear strain rate $\dot{\gamma}$ to an applied shear stress $\tau$:

$$\tau = \eta\dot{\gamma} \tag{6.1}$$

Many polymers obey a similar non-Newtonian power law relationship in which $\dot{\gamma}$ is raised to a power $m$ less than one. The viscosity is temperature dependent, decreasing with increasing temperature. For both polymers and liquid metals, a general relationship is

$$\eta = \eta_0 \exp\left(\frac{Q}{RT}\right) \tag{6.2}$$

where $\eta_0$ is a constant, $Q$ is the activation energy for flow, $R$ is the universal gas constant and $T$ is absolute temperature.

Low viscosity is generally desirable for RM since this means that the material will flow easily. This is important for FDM and SLS since powder binding is critical to successful processing. Low viscosity of binders is equally crucial for 3DP and other inkjet processes since droplet formation characteristics are strongly dependent on viscosity. Binders in SLS must also have a low melt viscosity to wet particles during the laser sweep. Finally, SLA (stereolithography apparatus) uncured resin viscosity must be controlled to ensure proper recoating between process layers. Viscosity is lowered significantly by increasing the temperature according to equation (6.2). The result is that most of the RM processes operate at elevated temperatures to maintain low viscosity. The viscosity of polymers varies significantly with temperature. The viscosity of metals ranges between about 0.2 mPa s for alkali metals to as high as 5 mPa s for $d$-transition metals. The viscosity of water at room temperature is 1 mPa s.

Solid particles added to a liquid increase the apparent viscosity. The viscosity as a function of solids loading $f_s$ may be written as [1]

$$\eta = \frac{\eta_a}{\left(1 - \frac{f_s}{f_{scr}}\right)^2} \tag{6.3}$$

where $\eta_a$ is the liquid viscosity with no solids and $f_{scr}$, which is approximately 0.62, is the critical solids loading at which all flow ceases. Figure 6.1 shows this behavior for a semi-solid Sn–15Pb alloy. The same behavior is observed in liquids containing foreign solid particulate. For RM processing of metals, solidification is usually associated with a temperature range in which solid and liquid coexist. It is desirable to process at temperatures above this slushy zone where the material is completely liquid and the viscosity is low. Other examples of solid–liquid mixtures in RM are binder–particle interactions during SLS and SLA using particle-charged resins. It is critical in post-processing infiltration of RM porous parts to maintain strong interparticle bonds; if this is not the case, the liquid infiltrant must not

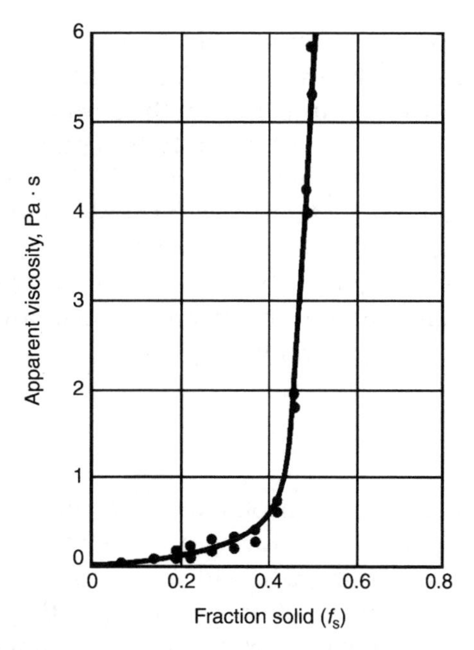


**Figure 6.1** Apparent viscosity as a function of solids loading for an Sn–15Pb alloy. (Reproduced from reference (2), Figure 15, p. 1405. With kind permission of Springer Science and Business Media)

exceed about 50 % or the part will slump and lose shape due to low apparent viscosity.

## 6.3 Photopolymerization

The enabling feature of stereolithography is the ability of uncured thermoset polymers to cross-link when exposed to light. Because few monomers absorb light in the near ultraviolet (UV) range, photosensitizers are added. These are chemicals that absorb light in the near UV range and dissociate into free radicals. Azoisopropane is a photosensitizer that is thermally stable over a wide range of temperatures and readily dissociates under UV light.

The rate of photopolymerization $R_p$ and the kinetic average chain length $v_0$ have been written as [3]

$$
\begin{aligned}
R_p &= -\frac{d[M]}{dt} = k_p[M][M^\bullet] = k_p[M]\sqrt{\frac{2.303[A]\phi P_0}{k_t}} \\
v_0 &= \frac{k_p[M]}{2\sqrt{2.303\,k_t\varepsilon[A]\phi P_0}}
\end{aligned}
\tag{6.4}
$$

where $[M]$ is the monomer concentration, $[M^\bullet]$ is the concentration of monomer radicals, $k_p$ is a polymerization rate constant, $[A]$ is the concentration of the photoinitiator, $\phi$ is the quantum yield (a number between 0 and 1 that represents the number of chains initiated per quantum of light), $P_0$ is the light source radiant power per unit area on the surface, $k_t$ is the rate constant for termination, which is orders of magnitude higher than those for initiation, and $\varepsilon$ is the molar absorptivity. For rapid photopolymerization, high concentrations of monomer, monomer radicals and photoinitiator are desired, and the light source should be intense and well coupled to the resin. High molecular weight or large values of $v_0$ are desired since low molecular weight often results in brittleness. Thus, laser power cannot be excessively high or resulting parts have poor mechanical properties.

## 6.4 Sintering

Sintering is the time-dependent consolidation of a porous medium. In RM this may occur during the process or during a post-process sintering treatment. Since virtually all properties improve with an increase in relative density, the desire is to minimize the porosity of finished articles. Because sintering is time dependent, the rate at which sintering occurs is important. Metals and ceramics sinter at rates much too slow for the phenomenon to occur during actual part build. However, polymers sinter at much higher rates, which may be consistent with rapid manufacturing processes. In any case, it is generally desired to produce final products as quickly as possible, whether the process step is direct RM processing or a pre- or post-processing treatment.

Viscous sintering, applicable to polymers, is described by neck growth in a two-particle system [4–6] (see Figure 6.2). The rate of growth of the neck $\dot{y}$ is given by Frenkel [4] as

$$\dot{y} = \frac{2}{3}\left(\frac{\Gamma}{\eta}\right)\frac{R}{y} \tag{6.5}$$

where $\Gamma$ is the surface tension, $\eta$ is the viscosity, $R$ is the particle radius and $y$ is the neck radius.

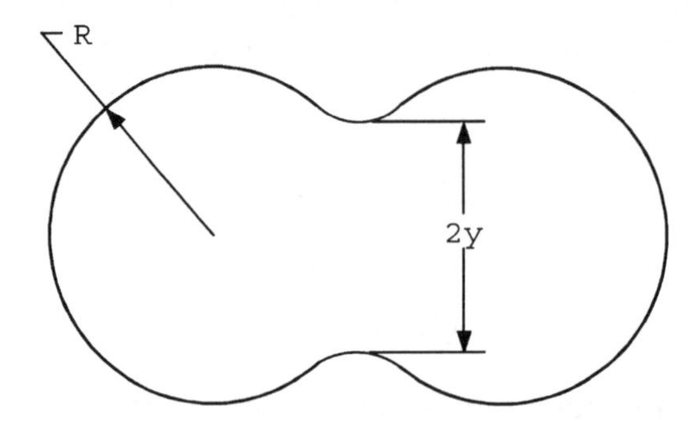


**Figure 6.2** Schematic of two-particle sintering

Scherer [5,6] described the densification rate for viscous sintering using a free strain term $e$, which may be related to the porosity $\varepsilon$. Based on a cubic array of cylinders, the change in free strain $e$ with time is written as

$$\begin{aligned} \frac{\partial e}{\partial t} &= -\frac{M}{\eta}\frac{(3\pi)^{\frac{1}{3}}}{6}\frac{2-3cx}{\sqrt[3]{x(1-cx)^2}}, \qquad \Delta \le 0.94 \\ \frac{\partial e}{\partial t} &= -\frac{M}{2\eta}\left(\frac{4\pi}{3}\right)^{\frac{1}{3}}\left(\frac{1}{\Delta}-1\right)^{\frac{2}{3}}, \qquad \Delta > 0.94 \end{aligned} \tag{6.6}$$

where $M = (\Gamma/R)[3/(4\pi)]^{\frac{1}{3}}$, $c = 8\sqrt{2}/(3\pi)$, $\eta$ is the viscosity, $\Delta$ is the relative density defined as the ratio of the part density to the material theoretical density, $\Gamma$ is the surface tension, $R$ is the particle initial radius and $x$ is defined as the ratio of the cylinder structure to the length of the cylinder. The porosity $\varepsilon = (1 - \Delta)$ is related to the free strain $e$ by

$$\begin{aligned} \varepsilon &= 1-(1-\varepsilon_0)\exp(-3e), \qquad \Delta \le 0.94 \\ \varepsilon &= 1-3\pi x^2+8\sqrt{2}x^3, \qquad \Delta > 0.94 \end{aligned} \tag{6.7}$$

where $\varepsilon_0$ is the initial porosity when $t = 0$.

Figure 6.3 shows the results of these relationships output as the sintering depth as a function of laser exposure time on an acrylonitrate butadiene styrene (ABS) powder bed.

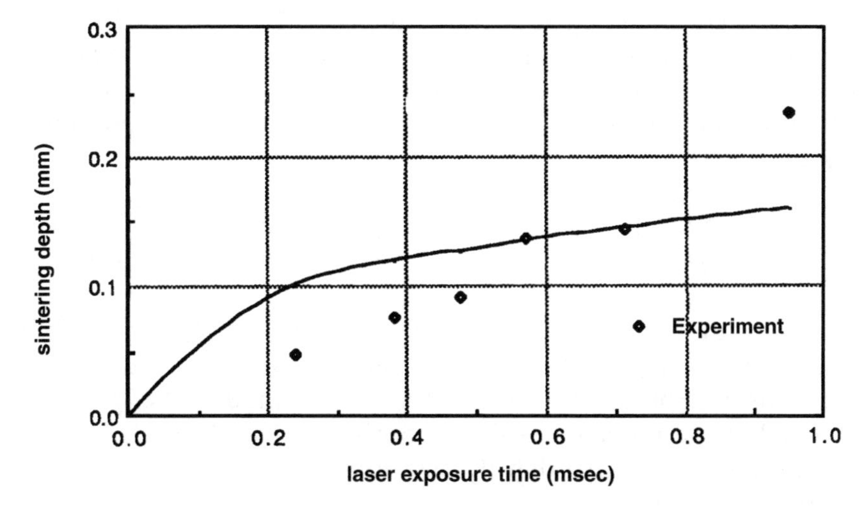


**Figure 6.3** Powder densification as a function of laser exposure time. ABS powder, heat input = $8.88 \times 10^7$ W m$^{-2}$, bed temperature = 20 °C. (Reproduced with permission from M.-s. M. Sun, J.J. Beaman and J.W. Barlow, Parametric analysis of the selective laser sintering process, in *SFF Symposium Proceedings* (eds J.J. Beaman, H.L. Marcus, D.L. Bourell and J.W. Barlow), Austin, Texas, 1990, pp. 146–54)

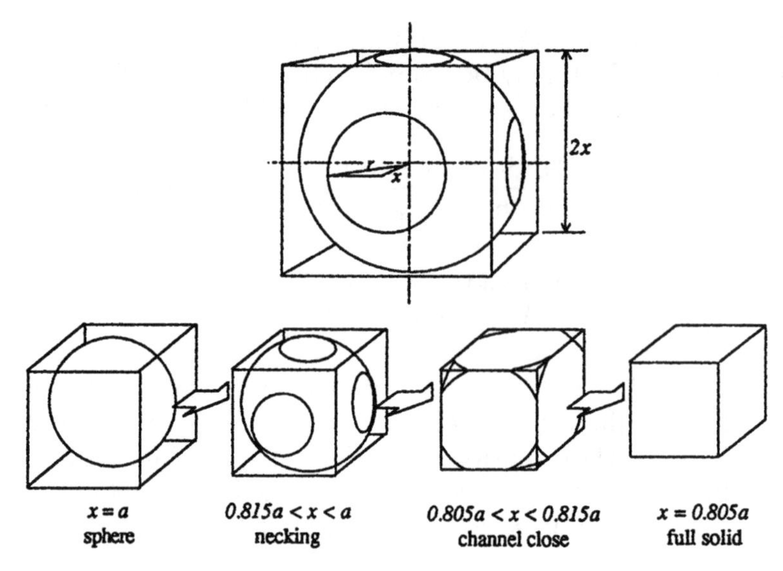


**Figure 6.4** Model for viscous sintering of spherical particles. (Reproduced with permission from M.-s. M. Sun, J.C. Nelson, J.J. Beaman and J.W. Barlow, A model for partial viscous sintering, in *SFF Symposium Proceedings* (eds H.L. Marcus, J.J. Beaman, J.W. Barlow, D.L. Bourell and R.H. Crawford), Austin, Texas, 1991, pp. 45–55)

Another viscous sintering approach [8] assumes particles in the form of spheres which sinter according to the Frenkel viscous sintering model (equation (6.5)). Each spherical particle of initial radius $a$ is effectively mushed into a shrinking cube of side dimension $2x$ (Figure 6.4).

During sintering, the cube dimension $2x$ decreases while the sphere radius $r$ increases to maintain constancy of volume. Pore separation occurs when $r = \sqrt{2}x$, associated with $x = 0.815\,a$. During this period, the sintering rate $\dot{x}$ is given by

$$\dot{x} = -\frac{\pi \Gamma a^2}{6\eta x^3}\left\{ r - (1-\zeta)x + \left[x - \left(\zeta + \frac{1}{3}\right)r\right]\frac{9(x^2 - r^2)}{18\,rx - 12\,r^2} \right\} \tag{6.8}$$

where $\zeta$ is a number between 0 and 1 associated with the probability of formation of sintering necks between particles and is a function of the relative density. For small values of $\zeta$ less than 0.3, the parameter is proportional to the relative density according to $\zeta \approx (2\Delta - 1)/3$. Figure 6.5 shows a plot of the relative density as a function of $\zeta$ and a dimensionless time factor $\pi\Gamma t/(6\,\mu a)$.

From the relationships described above, some general trends may be seen. Firstly, for increased flow and process speed, fine polymer particles at high temperature are desired. Particle size must be balanced against material cost and flow characteristics if spreading is necessary (e.g. SLS),

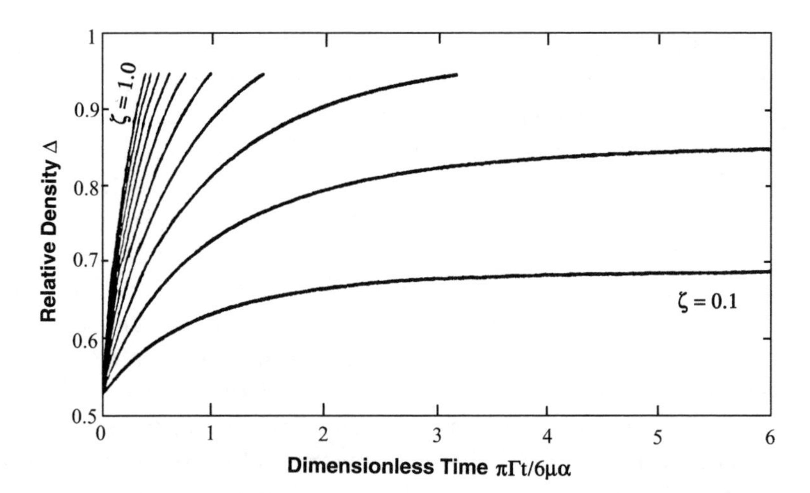


**Figure 6.5** Relative density of viscous sintered spherical particles as a function of dimensionless time $\pi\Gamma t/(6\mu a)$. (Reproduced with permission from M.-s. M. Sun, J.C. Nelson, J.J. Beaman and J.W. Barlow, A model for partial viscous sintering, in *SFF Symposium Proceedings* (eds H.L. Marcus, J.J. Beaman, J.W. Barlow, D.L. Bourell and R.H. Crawford), Austin, Texas, 1991, pp. 45–55)

since fine particulate does not spread as well as coarse powder. In SLS, temperature in polymer beds is usually set at a high value where no flow occurs. The laser sweep provides sufficient thermal energy to increase the temperature such that rapid flow occurs.

Solid-state sintering of metals and ceramics is limited to pre- and post-processing since the times required to accomplish significant densification are long. Rapid manufacturing of metals using processes involving powder spray and direct SLS provide high energy for melting and flow of liquid metal rather than solid-state sintering described here. For pressureless sintering, a reduction in total surface area is the only driving force for the densification of the powder mass. In the case of hot isostatic pressing (HIP), an additional driving force is also active, stemming from the stress created by the compaction pressure. To provide a convenient and unified form of the two driving forces, the surface area component of the driving force is converted to an effective stress and summed with the net applied HIP gas pressure, as shown in the following equations. A complete derivation of the equations is given by Ashby [9].

$$P_{\text{total}} = P_{\text{ext}} - P_0 + 3\,\Delta^2 \frac{\Gamma}{R}\left(\frac{2\,\Delta - \Delta_0}{1 - \Delta_0}\right) \qquad \text{[Stage I, open porosity]}$$

$$P_{\text{total}} = P_{\text{ext}} - P_{\text{int}}\frac{(1 - \Delta_{\text{c}})\Delta}{(1 - \Delta)\Delta_{\text{c}}} + 2\frac{\Gamma}{R}\left(\frac{6\,\Delta}{1 - \Delta}\right)^{\frac{1}{3}} \qquad \text{[Stage II, closed porosity]} \tag{6.9}$$

where $P_{total}$ is the effective applied driving pressure, $P_{ext}$ is the HIP pressure, $P_0$ is the initial pore pressure, $P_{int}$ is the internal pore pressure at pore close-off ( Stage I to Stage II transition), $\Delta$ is the relative density, $\Gamma$ is the powder surface energy, $R$ is the powder radius, $\Delta_0$ is the initial relative density and $\Delta_c$ is the relative density at pore closure/isolation. In practice, the contribution due to the surface energy reduction term is important in conventional, pressureless sintering, but it is minor relative to realistic HIP pressures. For example, using fairly typical values of $R = 25\,\mu m$ and $\Gamma = 2000\,mN\,m^{-1}$, it is seen that even near the end of the densification process ($\Delta = 0.97$), where the surface energy contribution is greatest, the effective pressure is still only 0.93 MPa. With typical HIP pressures on the order of 100 MPa, it is apparent that there is a substantial driving force increase associated with HIP compaction relative to conventional pressureless sintering.

In all stages of sintering, several densification mechanisms act simultaneously. In Ashby's initial analysis for pressure sintering maps, four independent mechanisms were included: lattice diffusion, grain boundary diffusion, power law creep (or dislocation creep) and plasticity (yielding/dislocation glide). In more recent work, the additional mechanisms of Nabarro–Herring creep and Coble creep were included for cases where the internal grain size of the particles is substantially smaller than the particle diameter [10]. Grain growth and grain boundary separation from pores during Stage II densification are also incorporated into the model.

By combining the driving force and densification mechanisms, it is possible to construct a sintering map, a plot of the relative density as a function of sintering temperature and time. An example is shown in Figure 6.6 for nickel alloy 625 powder selective laser sintered to $\Delta = 0.6$ and hot isostatic pressed for 1 or 8 hours at temperatures between 800 and 1000 °C. The computed isochronal lines show excellent agreement with experimental results. Also shown are the dominant sintering mechanisms.

Sintering maps can also be used to aid in powder pre-processing. An example is pre-processing of nanocrystalline zirconia prior to SLS [11]. Nanocrystalline ($D = 24\,nm$) zirconia was available in polymer-bound, agglomerated form (50 μm). This was unacceptable for SLS processing because the binder would melt and flow in an unpredictable fashion during SLS, and the coarse particles would not hold together. The need then was to determine a thermal excursion by which the nanocrystalline particulate within agglomerates might be debound and sintered together to high density, but for which the 50 μm agglomerates would not sinter.

Pressureless temperature densification maps for zirconia with a particle size of 24 nm and 50 μm appear as Figures 6.7 and 6.8 respectvely. It is observed that zirconia fine particles sinter lightly in 2 hours at 1050 °C and almost completely densify in 2 hours at 1250 °C. At 1250 °C, the dominant densification mechanism is boundary diffusion initially which transitions

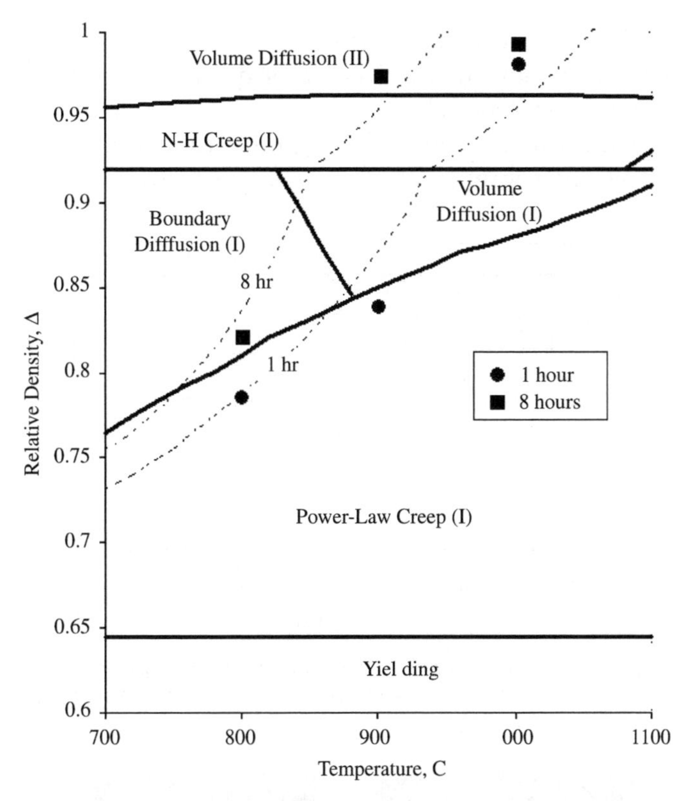


**Figure 6.6** Densification map for nickel alloy 625 powder after selective laser sintering. The particle diameter was 25 μm. (Reproduced from reference (11) with permission from TMS, Warrendale, Pennsylvania © 2000)

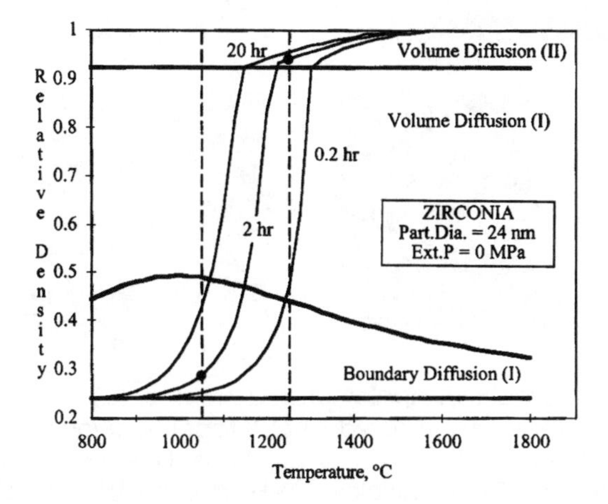


**Figure 6.7** Pressureless densification map for 24 nm zirconia powder. Densification occurs between 1050 and 1250 °C. (Reproduced from reference (11) with permission from TMS, Warrendale, Pennsylvania © 2000)

to volume (bulk) diffusion at about 45 % relative density. Further, from Figure 6.8, it is seen that 50 μm zirconia is quite resistant to pressureless densification to at least 2000 °C.

Figures 6.9(a) and (b) show low and high magnification scanning electron microscopy (SEM) micrographs of yttria-stabilized zirconia powder held for

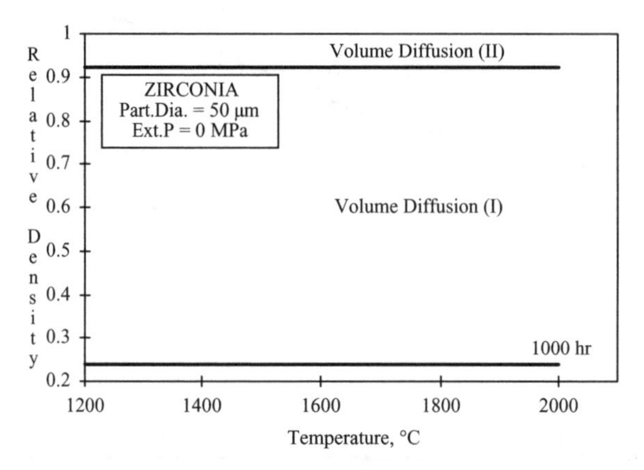


**Figure 6.8** Pressureless densification map for 50 μm zirconia powder. No densification is observed up to 2000 °C. (Reproduced from reference (11) with permission from TMS, Warrendale, Pennsylvania © 2000)

**Figure 6.9** SEM micrographs of agglomerated zirconia. (a) Held in air at 1050 °C for 2 hours. The agglomerated 50 μm powder holds together even though the binder has been burned off. (b) Held in air at 1050 °C for 2 hours. Fine, low-density nanocrystalline structure present within 50 μm particles. (c) Held in air at 1250 °C for 2 hours. The agglomerated structure remains. (d) Held in air at 1250 °C for 2 hours. Within 50 μm particles, the nanocrystalline particulate has fully densified. (Reproduced from reference (11) with permission from TMS, Warrendale, Pennsylvania © 2000)

2 hours in air at 1050 °C. In Figure 6.9(b), the fine particulate is visible and has lightly sintered. Figures 6.9(c) and (d) are SEM micrographs of powder held for 2 hours at 1250 °C. The coarse particles are still unsintered, but the fine-particle interiors are completely densified. This is consistent with the densification map observations.

## 6.5 Infiltration

Infiltration is a technique for increasing the end-use properties of RM parts. It is viable when the part has open porosity, which is porosity that forms a continuous tunnel-like network throughout the part. Since most properties are diminished by porosity (next section), it is generally desirable to eliminate as much porosity as possible. Several RM techniques, such as fused metal deposition create full-density parts directly. Others, such as 3DP (three-dimensional printing) and SLS, generally produce intermediate parts with open porosity. In traditional powder processing, hot pressing and other deformation processes are employed. These are of limited value in RM since the geometry is compromised by these techniques. High-temperature post-process sintering is sometimes useful if the kinetics for densification are acceptable (previous section). In some cases, it is preferred to infiltrate the part with a liquid material of lower melting point than the parent material. This results in a composite part with nominally full density. A key advantage over other full-density processing techniques is the ability to largely preserve part geometry, important for RM. Here, infiltration is usually a post-processing step wherein the porous part is heated in contact with the infiltrant to a temperature at which the infiltrant is molten and will wet the part. The infiltrant then soaks into the part similar to a sponge taking in water. Upon cooling, the infiltrant solidifies to produce the final part.

It is not possible in every case for a given infiltrant to soak or wick into a given porous preform. In many cases, spontaneous infiltration does not occur. Clearly, understanding of the fundamental process and material variable of the preform and the infiltrant are critical to successful infiltrated part production. The key consideration is the ability of the infiltrant to wet the solid preform. The wetting angle $\theta$ is related to the various surface energies involved according to Young's equation [12]:

$$\cos\theta = \frac{\gamma_{sv} - \gamma_{ls}}{\gamma_{lv}} \tag{6.10}$$

where $\gamma_{sv}$ is the surface energy associated with the solid surface being wet, $\gamma_{ls}$ is the surface energy associated with the liquid–solid interface and $\gamma_{lv}$ is the surface energy associated with the liquid infiltrant. Low wetting angles promote wetting.

For metals, the liquid surface energy $\gamma_{lv}$ varies from about 300 mN m$^{-1}$ for alkali metals and semi-metals to as high as 1800 mN m$^{-1}$ for $d$-transition metals [13]. The liquid surface energy may be measured experimentally using several test methods including the sessile drop test [14] and the pendant drop test [15]. Solid metal surface energies $\gamma_{sv}$ are theoretically 9 % higher than the metal liquid surface energy at the melting point, although actual values range between 4 and 33 % higher [16]. The solid surface energy is a strong function of cleanliness which motivates stringent surface preparation processes prior to joining in the soldering/brazing industry. Liquid–solid surface energies $\gamma_{ls}$ vary widely depending on the materials involved. It may be measured experimentally using the wetting balance test [17].

Infiltration as a post-processing step involves the introduction of a molten phase into a porous medium. The capillary force of liquid drawing into a pore structure of average pore diameter $d$ is equivalent to the application of an external pressure $P$ on the liquid in the pore structure described by the Washburn equation

$$P = -\frac{4\gamma_{lv}\cos\theta}{d} \tag{6.11}$$

A threshold pressure $P^*$ for spontaneous infiltration into particulate media is given as [18,19]

$$P^* = \frac{6\lambda(-\gamma_{lv}\cos\theta)\Delta}{(1-\Delta)D} \tag{6.12}$$

where $D$ is the particle size and $\lambda$ is the ratio of the particle actual surface area to the surface area of a sphere of the same volume. When the wetting angle for either equation (6.11) or (6.12) is less than 90°, wetting occurs and some infiltration will take place. The extent of infiltration depends on the magnitude of the effective pressure which is impacted not only by the degree of wetting but also other factors in the equations.

If infiltration works against gravity, an equilibrium height exists that is associated with the balance between the wetting and gravitational forces [20]:

$$h_{max} = \frac{4\gamma_{lv}\cos\theta}{dg\rho} \tag{6.13}$$

where $g$ is the gravitational acceleration and $\rho$ is the liquid density. The extent to which liquid will penetrate in a reasonable amount of time depends on the pore diameter, the liquid surface tension $\gamma_{lv}$ and also the liquid viscosity $\eta$. An empirical rule developed for infiltration of solder into a

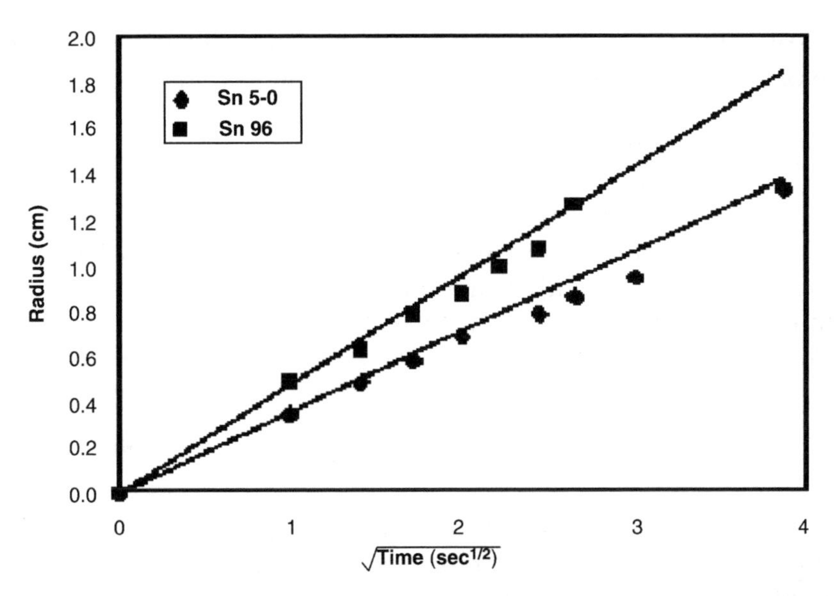


**Figure 6.10** Spreading of Pb–Sn solder into a thin copper powder bed. (Reproduced with permission from D.E. Bunnell, S. Das, D.L. Bourell, J.J. Beaman and H.L. Marcus, Fundamentals of liquid phase sintering during selective laser sintering, in *SFF Symposium Proceedings* (eds H.L. Marcus, J.J. Beaman, D.L. Bourell, J.W. Barlow and R.H. Crawford), Austin, Texas, 1995, pp. 440–7)

copper powder bed [21] defines the radius of spreading $R$ as proportional to $\sqrt{t/D}$, where $D$ is the particle size of the powder mass being infiltrated and $t$ is the time elapsed after infiltration commences. Figure 6.10 shows the correlation of the empirical relationship to actual measured spread envelopes for two compositions of lead–tin solder into copper powder.

Figure 6.11 is a photograph of a silicon-infiltrated silicon carbide part produced by indirect selective laser sintering of silicon carbide followed

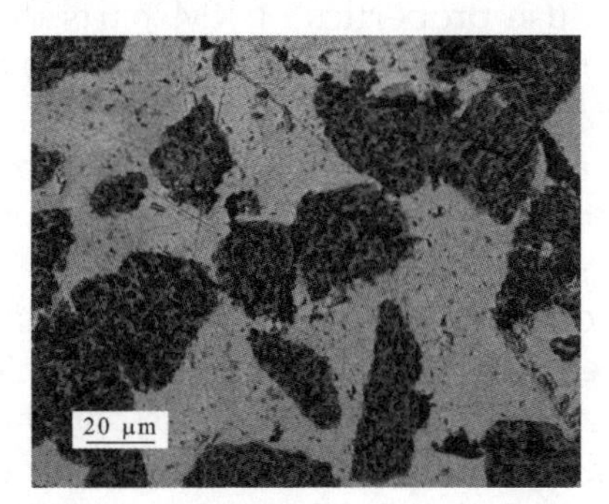


**Figure 6.11** A silicon-infiltrated silicon carbide preform produced using selective laser sintering. (a) The part is 5 cm by 6 cm in cross-section. (b) A micrograph of a cross-section showing dark SiC and light Si infiltrant. (Courtesy of R.S. Evans, S. Barrow and D.L. Bourell)

by post-process pressureless infiltration with silicon. The threshold pressure for the onset of infiltration according to equation (6.12) is

$$P^* = \frac{6\lambda(-\gamma_{lv}\cos\theta)\Delta}{(1-\Delta)D} = \frac{6(1.3)(-846\,\text{mJ}\,\text{m}^{-2}\cos 41^\circ)0.4}{(1-0.4)40\,\mu\text{m}} = -83\,\text{kPa} \qquad (6.14)$$

where $\gamma_{lv} = 885\,\text{mJ}\,\text{m}^{-2} - 0.28(\text{mJ}\,\text{m}^{-2})\text{K}^{-1}(T - 1683\,\text{K})$, $\Delta = 0.4$, $\theta = 41^\circ$ and $D = 40\,\mu\text{m}$ [22]. This negative pressure is conceptually a vacuum on the molten metal offsetting capillarity forces for infiltration into the pore structure. Alternatively, the Washburn pressure comes from equation (6.11):

$$P = -\frac{4\gamma_{lv}\cos\theta}{d} = -\frac{4(-846\,\text{mJ}\,\text{m}^{-2}\cos 41^\circ)}{40\,\mu\text{m}} = -62\,\text{kPa} \qquad (6.15)$$

where $d \approx 40\,\mu\text{m}$ [22]. Pressureless infiltration may work against gravity, reaching an equilibrium height defined according to equation (6.13). For the silicon-infiltrated silicon carbide, this maximum height is

$$h_{\max} = \frac{4\gamma_{lv}\cos\theta}{dg\rho} = \frac{4(-846\,\text{mJ}\,\text{m}^{-2}\cos 41^\circ)}{(40\,\mu\text{m})(9.8\,\text{m}\,\text{s}^{-2})(2.65\,\text{g}\,\text{cm}^{-3})} = 2.5\,\text{m} \qquad (6.16)$$

This compares well with other research, in which a maximum height of about 2 m was calculated [20]. Since the effective height is over 2 m, the driving force for pressureless infiltration is large for most small parts, and infiltration should be effectively achieved.

## 6.6 Mechanical Properties of RM Parts

The in-use properties of RM parts define the scope of applicability for the technology. Parts that are fully dense and homogeneous manifest properties associated with their microstructure. For a number of direct processes, the parts are a solidified structure of polymer or metal. The associated cast properties are generally obtained. For example, Ti–6Al–4V was direct selective laser sintered to create a nominally fully dense part [24]. Table 6.1 lists the properties and structure information for wrought, cast and selective laser sintered material as well as information on the impurities in PREP and Ar atomized powder, precursors for the SLS processing. It is seen that the mechanical properties and impurity concentrations are comparable to cast and annealed wrought materials.

In the case of polymer mechanical properties, layer manufactured parts usually fall short of their injection molded counterparts. This is due to issues such as porosity and the absence of high pressures and consolidation during

**Table 6.1** Properties of Ti-6Al-4V in cast, wrought, SLS and powder form

| Processing | References | Hardness HRC | Tensile strength (MPa) | Elongation (%) | Oxygen (%) | Nitrogen (%) |
|---|---|---|---|---|---|---|
| Cast Grade C-5 | [25] | 39 | 895 | 6 | 0.25 | 0.05 |
| Cast Grade C-6 | [25] | 36 | 795 | 8 | 0.20 | 0.05 |
| Annealed wrought Grade 5 | [26] | — | 895 | 10 | 0.20 | 0.05 |
| PM HIP | [24] | 34–36.5 | — | — | — | — |
| SLS direct (Ar atomized powder) | [24] | 36 | 1120 | 5 | 0.23 | 0.037 |
| PREP powder | [24] | — | — | — | 0.19 | 0.01 |
| Ar atomized | [24] | — | — | — | 0.196 | 0.02 |

**Table 6.2** Tensile properties of conventionally processed and selective laser sintered nylon

| Material/process | UTS (MPa) | $E$ (GPa) | Elongation at break (%) |
|---|---|---|---|
| Nylons/cast/molded [27] | 55–83 | 1.4–2.8 | 60–200 |
| Nylon-12/selective laser sintered [28] | 46 | 1.8 | 12 |

manufacturing. Table 6.2 compares tensile properties of laser sintered nylon with the range of tensile properties found in cast and molded nylons. While the stiffness of selective laser sintered parts is comparable with other processes, the strength and ductility fall short. ABS parts made by fused deposition modeling also fail to match the mechanical properties of molded parts, especially in the $Z$ direction.

This section describes the effects of residual porosity for a variety of end-use properties. For many applications, mechanical properties are important if not critical. Residual porosity negatively impacts all the mechanical properties, in order of decreasing severity: fracture, fatigue, strength, ductility, modulus. The effects of relative porosity $\varepsilon$ on each is described.

For strength, the general dependence takes the form [29]

$$\sigma = K\sigma_0(1-\varepsilon)^m = K\sigma_0(\Delta)^m \tag{6.17}$$

where $\sigma$ is the strength, $K$ and $m$ are constants and $\sigma_0$ is the wrought strength of the same alloy. $K$ is geometry and processing dependent. The relationship is generally applied to yield strength, tensile strength and three- or four-point bend strength. Figure 6.12 is a logarithmic plot of the tensile strength of SLS processed bronze–nickel powder. The linear relationship for as-SLS processed and post-process liquid phase sintered parts confirms the effect of porosity on strength.

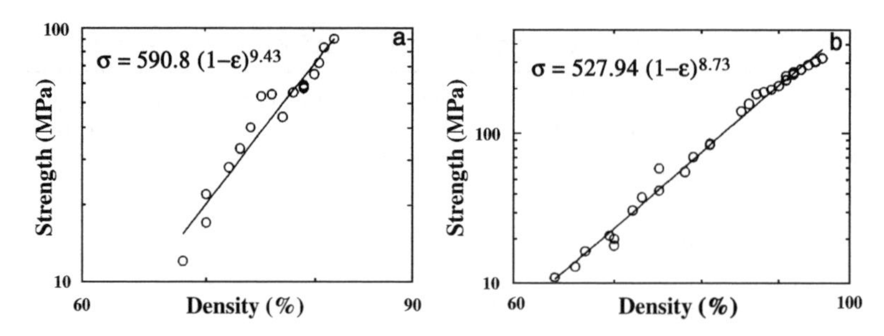


**Figure 6.12** Room-temperature tensile strength of pre-mixed SLS (90Cu–10Sn) bronze and commercially pure nickel powder as a function of relative density $\Delta = 1 - \varepsilon$. (a) As SLS processed. (b) SLS processed and sintered at 900–1100 °C for 1 to 10 hours. (Reproduced with permission from M.K. Agarwala, D.L. Bourell, B. Wu and J.J. Beaman, an evaluation of the mechanical behavior of bronze–Ni composites produced by selective laser sintering, in *SFF Symposium Proceedings* (eds H.L. Marcus, J.J. Beaman, J.W. Barlow, D.L. Bourell and R.H. Crawford), Austin, Texas, 1993, pp. 193–203)

Ductility of parts with residual porosity may be predicted using the relationship [31]

$$Z = \frac{(1-\varepsilon)^{\frac{3}{2}}}{(1+C\varepsilon^2)^{\frac{1}{2}}} \tag{6.18}$$

where $Z$ is the ratio of the porous material ductility to equivalently processed full density ductility and $C$ is an empirical constant related to the sensitivity of the ductility to the presence of pores. Large values of $C$ correspond to highly sensitive ductility. For example, a residual porosity of 10 % causes a 20 % reduction in ductility if $C$ equals 10, but it causes a 97 % reduction if $C$ equals 100 000. Generally, $C$ varies between about 100 and 100 000. Figure 6.13 shows the effect of residual porosity on the elongation to fracture of two conventionally sintered iron products.

Fatigue and fracture behavior of RM parts are strongly influenced by residual porosity [32]. Not only is the volume fraction of pores important but so also are the spacing between pores, the average pore size and the morphology of the pores, particularly ones that reside on the surface. Pores generally increase the threshold stress intensity for crack initiation but lower the resistance to crack propagation.

For many materials with low residual porosity, the fatigue endurance limit is about 35 % of the tensile strength [32], compared to about 50 % for bulk materials. The stress field around an isolated spherical pore is about twice the far-field stress level.

The plane strain fracture toughness $K_{IC}$ has been shown to be a strong function of porosity [32]. For quenched and tempered steels, the fracture toughness decreases by about 100 MPa$\sqrt{\text{m}}$ for each per cent of porosity.

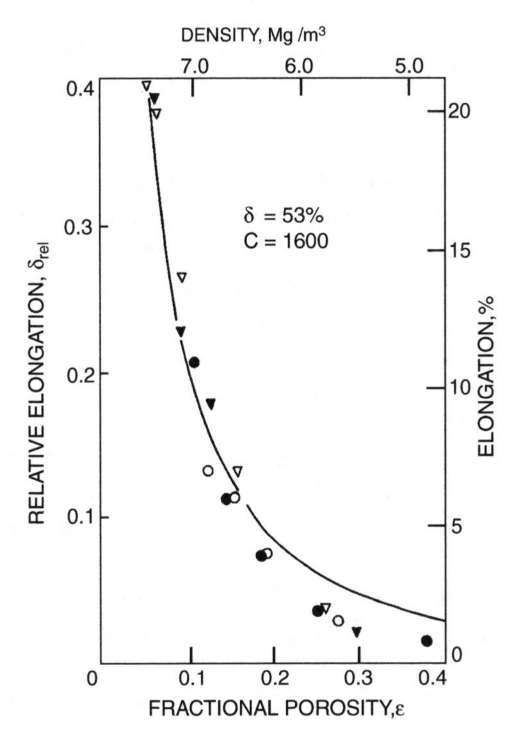


**Figure 6.13** Relative ductility as a function of fractional porosity for pure iron. Various particle sizes and purity. (Reproduced with permission from R. Haynes, A study of the effect of porosity content on the ductility of sintered metals, *Powder Metallurgy*, 1977, **20**, 17–20, Figure 3)

## 6.7 Materials for RM Processes

Numerous laboratories around the world have researched and developed materials for various RM processes. This work continues, and new materials and processes are being disseminated, discussed and brought to market. Below is a list of commercially available materials systems from a selected list of current manufacturers. Together they include photocuring resins, viscous-binder polymers, infiltrated metal, direct metal and infiltrated non-metallics.

1. *Stereolithography*. All commercial photopolymers for SLA are proprietary epoxies and acrylate–epoxy hybrids. 3D Systems markets four photopolymers currently. The first is a low-ash material for casting patterns. The other three photopolymers offer long vat life, optical clarity, smooth side walls, tack-free down faces, thin cured line width, low viscosity and low curing shrinkage.
2. *Selective laser sintering*. Commercial materials are polymers, metal or ceramic binder systems and direct metal systems. The most popular

material is polyamide, available in neat and glass-reinforced formulations. An aluminum-filled polyamide has been developed by EOS GmbH. Polystyrene and a resin-coated sand are available as well. 3D Systems offers three commercial SLS powders. First is a nylon formulated with or without glass reinforcement. Also available is an AISI-SAE A6 tool steel material designed to be infiltrated with metal after SLS. Another offering is a poly(methylmethacrylate) (PMMA) material with a low-ash content that produces porous parts designed to be infiltrated with foundry wax prior to employment as a casting pattern.

3. *Fused deposition modeling*. Stratasys has developed extrusion filaments of ABS, polycarbonate and polyphenylsulfone.
4. *Fused metal deposition* (LENS, DLM, direct SLS, POM). A variety of metal powders have been processed using direct powder techniques, including tool steel (A2, H13, H19, P20, P21, S7), stainless steel (304, 316, 420, 15-5PH, 17-4PH), nickel alloys (IN625, IN718, Hastelloy X), cobalt alloys (Stellite 6, Stellite 21, Stellite 706), aluminum alloy (4047), copper alloys and titanium alloys (CP Ti, Ti–6Al–4V, Ti–6Al–2Sn–4Zr–6Mo).
5. *Three-dimensional printing*. Z-Corp offers a plaster–ceramic composite as well as a starch-based material and a straight plaster. Infiltrants include epoxy, urethane cyanoacrylates and a low melting point wax.

## 6.8 The Future of Materials in Rapid Manufacturing

Materials developments in freeform fabrication have from the beginning powered enabling technological advance. Stereolithography hinges on photopolymers. Control of the thermal characteristics of polymers is crucial in fused deposition modeling. Powder production and morphology are significant in the advance of selective laser sintering.

As freeform fabrication matured in the 1990s, materials and processes have become more sophisticated. For example, in SLS, the first materials were thermoplastics processed in 1986. The materials demands were relatively elementary, using a low melting point plastic that will melt under the laser and re-solidify to produce a part. Within three years, the first direct metal system was investigated, with copper powder mixed with either lead or tin. From here, improvements in process and atmosphere control enabled direct SLS of ferrous materials, titanium and nickel-based alloys. The need for production of functional parts in commercial machines spurred development of metal-fusible binder systems coupled with post-processing binder burn-out and metal infiltration. Steel skeletons infiltrated with copper alloys were commercialized in the 1990s. Recent research focuses on development of ceramic composites similarly processed and infiltrated with higher melting point materials like silicon. Successful RM of these advanced materials systems requires understanding of laser-binder coupling, binder flow and dissociation characteristics, particle size effects, processing parameter optimization, infiltration kinetics and chemical interactions, as well as powder production and economic considerations.

Materials sophistication will continue to advance in the future. As societal demands develop for functional parts and devices with engineered properties and complex geometry, research and development will continue to satisfy those demands. Areas where RM is expected to make an impact include micro- and possibly nano-manufacturing. New freeform processes will emerge based on microelectronics photolithography and microelectromechanical systems (MEMS). Continued development of materials that are biocompatible will enhance successful integration of RM in biomedical applications. Even now, elementary research is being conducted worldwide to deposit and nurture living matter using freeform technologies. Mass customization will continue to mature as RM bears on consumer needs. Current examples include manufacturing of 'invisible' orthodontics and hearing aid shells, although possibilities are being demonstrated in areas such as textiles and implants. Future trends in materials development will follow the time-tested motto of RP/RM: 'better, faster, cheaper'.

## Acknowledgement

The author is indebted to Neil Hopkinson of Loughborough University for his input on polymer mechanical properties.

## References

1. German, R.M. (1990) Supersolidus liquid phase sintering II: densification theory, *International Journal of Powder Metallurgy*, **26**(1), 35–43.
2. Joly, P.A. and Mehrabian, R. (1976) The rheology of a partially solid alloy, *Journal of Materials Science*, **11**, 1393–1418.
3. Beaman, J.J., Barlow, J.W., Bourell, D.L., Crawford, R.H., Marcus, H.L., and McAlea, K.P. (1997) *Solid Freeform Fabrication: A New Direction in Manufacturing*, Kluwer Academic, Norwell, Massachusetts, p. 104.
4. Frenkel, J. (1945) Viscous flow of crystalline bodies under the action of surface tension, *Journal of Physics (USSR)*, **9**, 385.
5. Scherer, G.W. (1986) Viscous flow under a uniaxial load, *Journal of American Ceramic Society*, **69**(9), 206–7.
6. Scherer, G.W. (1986) Sintering of low density glasses: I–Theory, *Journal of American Ceramic Society*, **60**(5–6), 236–9.
7. Sun, M.-s. M., Beaman, J.J. and Barlow, J.W. (1990) Parametric analysis of the selective laser sintering process, in *SFF Symposium Proceedings*,

(eds J.J. Beaman, H.L. Marcus, D.L. Bourell and J.W. Barlow), Austin, Texas, pp. 146–54.

8. Sun, M.-s. M., Nelson, J.C., Beaman, J.J. and Barlow, J.W. (1991) A model for partial viscous sintering, in *SFF Symposium Proceedings*, (eds H.L. Marcus, J.J. Beaman, J.W. Barlow, D.L. Bourell and R.H. Crawford), Austin, Texas, pp. 46–55.
9. Ashby, M.F. (1990) *Sintering and Isostatic Pressing Diagrams*, Published by M.F. Ashby, Department of Engineering Cambridge.
10. Helle, A.S., Easterling, K.E. and Ashby, M.F. (1985) Hot-isostatic pressing diagrams: new developments, *Acta Metallurgica*, **33**(12), 2163.
11. Bourell, D.L., Wohlert, M. and Harlan, N. (2000) Powder densification maps and applications in selective laser sintering, in *Deformation, Processing and Properties of Structural Materials – A Symposium Honoring Oleg D. Sherby* (eds E.M. Taleff, C.K. Syn and D.R. Lesuer), Warrendale, Pennsylvania, TMS, pp. 219–30.
12. Heady, R.B. and Cahn, J.W. (1970) An analysis of the capillary forces in liquid-phase sintering of spherical particles, *Metallurgical Transactions*, **1**(1), 185–9.
13. Lide, D.R. (Editor-in-Chief) (1991) *CRC Handbook of Chemistry and Physics*, 72nd edn, CRC Press, Boca Raton, Florida, p. 4.136.
14. Iida, T. and Guthrie R.I.L. (1988) *The Physical Properties of Liquid Metals*, Clarendon Press, Oxford.
15. Bunnell, D.E., Bourell, D.L. and Marcus, H.L. (1994) Fundamentals of liquid phase sintering related to selective laser sintering, in *SFF Symposium Proceedings* (eds H.L. Marcus, J.J. Beaman, J.W. Barlow, D.L. Bourell and R.H. Crawford), Austin, Texas, pp. 379–86.
16. Brophy, J.H., Rose, R.M. and Wulff, J. (1964) *The Structure and Properties of Materials*: Vol. II, *Thermodynamics of Structure*, John Wiley & Sons, Inc., New York, pp. 49–52.
17. Vianco, P.T., Hosking, F.M. and Rejent, J.A. Solderability testing of Kovar with 60Sn–40Pb solder and organic fluxes, *Welding Journal*, 1990, **6**, 230s–40s.
18. Mortensen, A. and Cornie, J.S. (1987) On the infiltration of metal matrix composites, *Metallurgical Transactions A*, **18A**, 1160–3.
19. Oh, S.-Y., Cornie, J.A. and Russell, K.C. (1989) Wetting of ceramic particulates with liquid aluminum alloys: Part I. Experimental techniques, *Metallurgical Transactions A*, **20A**, 527–32.
20. Gern, F.H. (1995) Interaction between capillary flow and macroscopic silicon concentration in liquid siliconized carbon/carbon, *Ceramic Transactions*, 1995, 58, 149.
21. Bunnell, D.E., Das, S., Bourell, D.L., Beaman, J.J. and Marcus, H.L. (1995) Fundamentals of liquid phase sintering during selective laser sintering, in

*SFF Symposium Proceedings* (eds H.L. Marcus, J.J. Beaman, D.L. Bourell, J.W. Barlow and R.H. Crawford), Austin, Texas, pp. 440–7.
22. Wang, H. (1999) Advanced processing methods for microelectronics industry silicon wafer handling components, PhD Dissertation, The University of Texas at Austin, Austin, Texas, pp. 37–44.
23. Evans, R.S., Barrow, S. and Bourell, D.L. (2004) Unpublished Research.
24. Das, S., Wohlert, M., Beaman, J.J. and Bourell, D.L. (1998) Processing of titanium net shapes by SLS/HIP, in *SFF Symposium Proceedings* (eds D.L. Bourell, J.J. Beaman, R.H. Crawford, H.L. Marcus and J.W. Barlow), Austin, Texas, pp. 469–77.
25. ASTM B367-93 (1993) *Standard Specification for Titanium and Titanium Castings*, American Society for Testing and Materials, Philadelphia, Pennsylvania.
26. ASTM B348-95a (1995) *Standard Specification for Titanium and Titanium Alloy Bars and Billets*, American Society for Testing of Materials, Philadelphia, Pennsylvania.
27. Kalpakjian, S. and Schmid, S. (2001) *Manufacturing Engineering and Technology*, 4th edn, Prentice-Hall, Englewood Cliffs, New Jersey.
28. Zarringhalam, H. and Hopkinson, N. (2003) Post-processing of Duraform$^{TM}$ parts for rapid manufacture, in Proceedings of the 14th Solid Freeform Fabrication (SFF) Symposium, Austin, Texas, 4–6 August 2003, pp. 596–606.
29. German, R.M. (1984) *Powder Metallurgy Science*, Metal Powder Industries Federation, Princeton, New Jersey.
30. Agarwala, M.K., Bourell, D.L., Wu, B. and Beaman, J.J. An evaluation of the mechanical behavior of Bronze–Ni composites produced by selective laser sintering, in *SFF Symposium Proceedings* (eds H.L. Marcus, J.J. Beaman, J.W. Barlow, D.L. Bourell and R.H. Crawford), Austin, Texas, pp. 193–203.
31. Haynes, R. (1977) A study of the effect of porosity content on the ductility of sintered metals, *Powder Metallurgy*, **20**, 17–20.
32. German, R.M. and Queeney, R.A. (1998) Fatigue and fracture control for powder metallurgy components, in *ASM Handbook*, Vol. 7, *Powder Metal Technologies and Applications*, 10th edn, ASM International, Materials Park, Ohio, pp. 975–64.

# 7

# Functionally Graded Materials

Poonjolai Erasenthiran
*Loughborough University*

Valter E Beal
*Universidade Federal De Santa Catarina*

## 7.1 Introduction

Functionally graded materials (FGMs) are a form of composite where the properties change gradually with position. The concept of graded composition is illustrated in Figure 7.1. The gradient can be tailored to meet specific needs through the utilisation of composite components. The change in the property of the material is caused by a position-dependent chemical composition, microstructure or atomic order [1].

One type of composite aims to reinforce material properties by mixing a dispersed phase homogeneously within the matrix. Another type of composite is characterized by having different material characteristics on separate surfaces or in separate parts. Typical examples are coatings to enhance the surface properties. However, the sharp boundary existing within the structure often exhibits various adverse effects on the material properties.

Most FGMs have a gradually varied composition between different constituents. This eliminates mismatch of material properties; thus thermal (residual) stresses in parts that possibly cause fatigue can be significantly decreased. At the same time, the properties of an FGM are engineered to fulfil the various requirements, as seen in Figure 7.2, where these

*Rapid Manufacturing: An Industrial Revolution for the Digital Age*
Editors N. Hopkinson, R.J.M. Hague and P.M. Dickens 

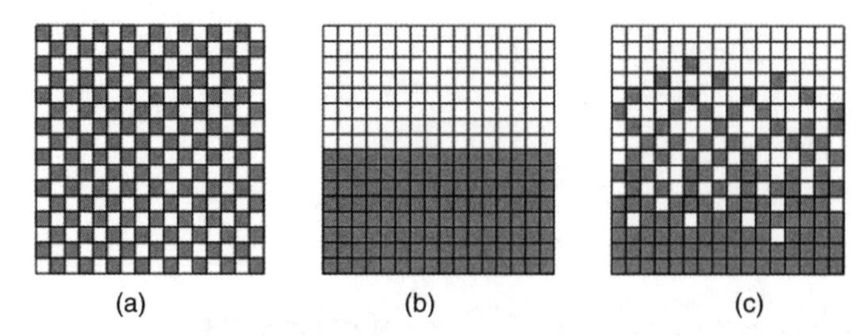


**Figure 7.1** Illustration of (a) homogeneous, (b) coated or joint type and (c) FGM

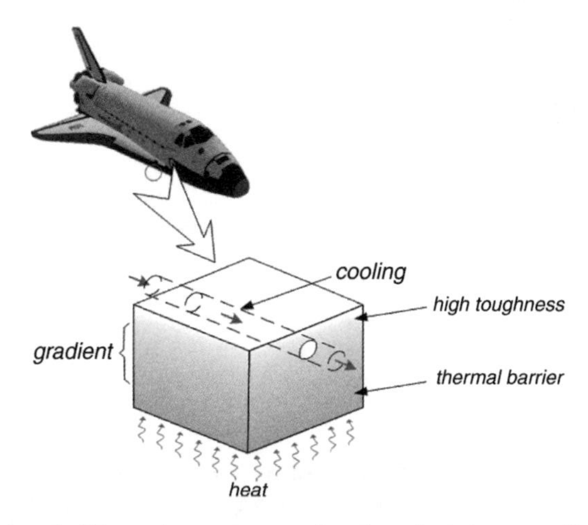


**Figure 7.2** Example of different requirements of material properties in different locations within a part

requirements are more difficult to meet completely through the approach of composite materials or conventional material processing methods.

The advantages of using FGM components were first recognized by Bever and Duwez [3] and Shen and Bever [4] in 1972. However, only in the early 1980s was systematic research on manufacturing processes for FGMs actively carried out in Japan and aimed at eliminating the microscopic boundary to improve the materials functions. In 1988 a national forum was established in Japan in recognition of the future of this technology. Subsequently, various techniques were established to process FGM components [5–8].

## 7.2 Processing Technologies

FGM components can be manufactured by various methods, as shown in Figure 7.3. However, at present, manufacturing of FGMs is based on

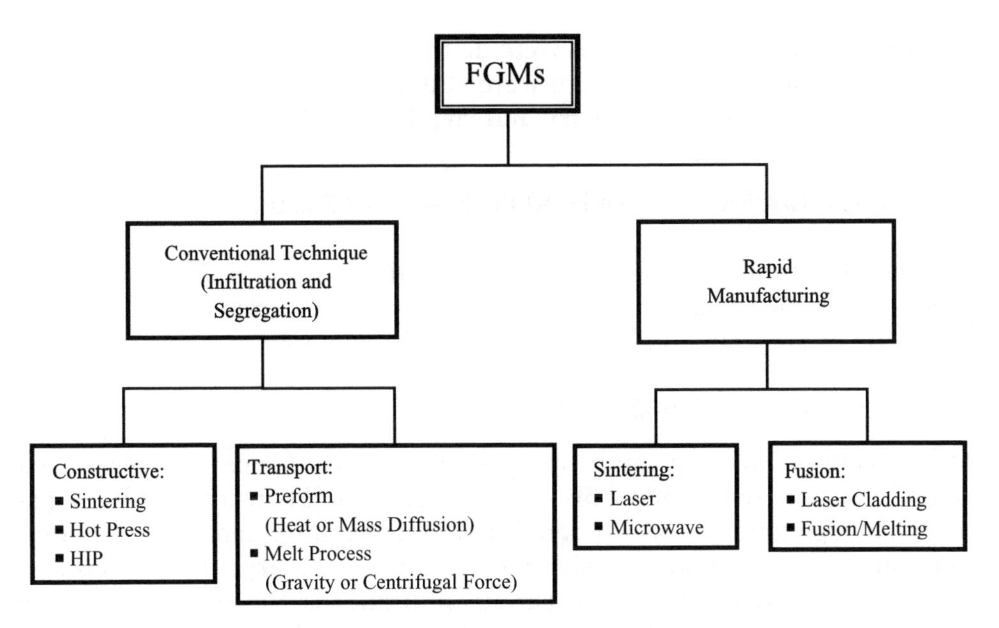


**Figure 7.3** Manufacturing methods of the FGMs components

constructive and transport-based processes [9]:

1. Constructive processes are based mainly on powder densification/sintering techniques and coating processes. This can be divided into two:
   - Bulk processing in which premixed powder is stacked in a stepwise or continuous manner according to the pre-designed and engineered spatial distribution of composition. The powder is then consolidated, where the options include sinter, hot press or hot isostatic pressing (HIP), etc.
   - Layer processing, where the materials are deposited on the surface by a laser (e.g. laser cladding), high pressure (thermal spraying), high voltage electrodes (sputtering), physical vapour deposition (PVD) or chemical vapour deposition (CVD).
2. Transport-based processes are those associated with mass transport, thermal process, settling and centrifugal processes and infiltration and macro-segregation processes.
   - Preform processing by means of heat or mass diffusion. The preforms can be either dense or porous. A simple example is to infiltrate a porous structure with another material. By applying thermal gradients or pressure, FGMs can be produced.
   - Melt processing, including settling under gravity or centrifugal force. It is difficult to control the compositional gradients.

Preforms and simple geometries were developed in the early 1990s [1] using these techniques. However, there were still limitations concerning material combinations, specimen geometry and cost [1].

## 7.3 Rapid Manufacturing of FGM Parts – Laser Fusion

At present various research are being carried out using RM techniques to produce complex shapes rapidly. This is generally produced using the technique of laser sintering or laser fusion of powders.

### *7.3.1 Liquid Phase Sintering (LPS)*

The concept of sintering can be categorised into two forms. One is focused on densification, where materials are processed to full density by sintering at relatively high temperature. Materials like silicon nitride, cemented carbides, steels and silicon carbide are sintered to fully dense structures for cutting tools, tips and many other industrial applications. The other form is to minimise the dimensional change and shows controlled porosity for filters, bearings and permeators, etc. [10].

Sintering temperatures are usually between 0.5 and 0.8 of the absolute melting temperature, when the process is commonly referred as solid-state sintering. The mass transport mechanisms, depending on the sintering stage, include surface diffusion, volume diffusion, grain boundary diffusion, viscous flow, plastic flow and even vapour transport from solid surfaces [10]. If the sintering temperature phase increases, liquid is generated and prevails during the sintering cycle (called LPS). The liquid improves the mass transport rates and also exerts a capillary pull on the particles to facilitate densification. The advantages of cost and shortened processing time mean that more than 70 % of sintered products are manufactured with the presence of a liquid phase [10].

The classic liquid phase sintering stages involve mixed powders which form a non-reactive liquid on heating. The wetting liquid provides a capillary force that pulls the solid particles together and induces particle rearrangement. In addition, the liquid gives rapid mass transport at the sintering temperature. This results in solution–reprecipitation and improved grain packing by grain shape accommodation [10]. Solution refers to solid dissolution into the liquid, while reprecipitation refers to solid leaving the liquid by precipitation on existing grains. Both processes occur simultaneously during liquid phase sintering (LPS). The final densification (solid skeleton sintering) is the final stage of LPS and is controlled by densification of the solid structure. The rigid skeleton of contacting solid grains makes densification slow and microstructural coarsening may happen. A schematic diagram of LPS is given in Figure 7.4.

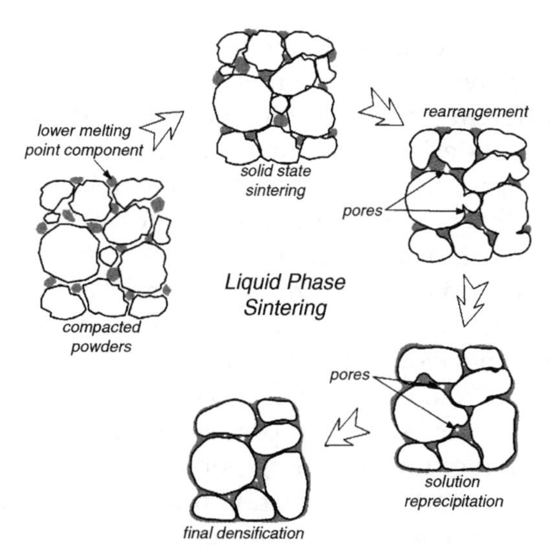


**Figure 7.4** Stages of liquid phase sintering (adapted from German (10))

### 7.3.2 LPS in Laser Processing Powders or FGMs

To create a dense structure of multiple materials through a laser-based process using powder deposition, there are a few approaches regarding the powder system. The first one is to coat the structural material with another material, usually of a lower melting point. An example is the indirect laser sintering in rapid tooling. The LaserForm™ powder produced by 3D Systems for sintering in Vanguard system (previously the DTM system), is coated material with poly(methylmethacrylate) (PMMA) binder (this is also discussed in Chapter 5 on Emerging Processes). Likewise, polystyrene and polyethyleneglycol-coated SS316 powder has been developed by the Fraunhofer Institute of Chemical Technology [11].

It is also possible to use pre-alloyed powder with a desired composition. Furthermore, there are many factors involved in the atomization of metals that make powdered materials behave differently from bulk material [12]. This may complicate the control of different chemical constituents in the powders. However, Kruth also showed that pre-alloyed powder (WC–Co) yielded higher selective laser sintering (SLS) green densities and strength as compared to processing of mixtures of individual powders. The resultant microstructure was also finer and exhibited less porosity with the pre-alloyed powder [13].

The third method is to mix separate components in the powder mixture. A commercial example is the low melting point mixture of bronze–nickel or bronze–stainless steel powders by EOS, Germany [14]. The major concerns include the degree of mixing due to the different density of each component and the various behaviours of different powders to the laser irradiation. Segregation of different phases in the powder mixture and in the fused sample may cause undesirable inhomogeneity of material properties. The advantage of this approach is that it is simple and able to achieve a high level of packing density by blending powders of different particle sizes.

Powder systems with a polymer binder or metals of low melting point have been generally used as metal prototypes or in batch production of parts in low volumes. Some structural materials, e.g. stainless steel or tool steel powders, have been developed for directly making parts or tools. It is clear that the production of dense structures directly by laser–powder deposition without any furnacing or infiltration procedure needs to melt the materials or at least the material of lower melting point by taking advantage of liquid phase sintering. Moreover, the interaction time of the laser and powders is usually less than one second and the powder rearrangement mainly relies on the melt and capillary penetration [15]. Despite the short time of powder densification, a benefit of laser–powder deposition is the limited grain growth at high temperatures due to rapid solidification rates. Such characteristic microstructures usually give materials a higher strength and hardness, which is an advantage for many industrial applications.

It is worth noting that there will be some thermodynamic enthalpy of mixing involved when some alloying reactions occur during laser melting. Depending on the endothermic or exothermic effect, this factor influences the homogeneity as well as the rate of solidification of the alloy and consequently the microstructure and properties of the deposit [15].

Research at the University of Texas at Austin has primarily focused on developing a modified Rapid Prototyping (RP) system to allow the fabrication of parts with multi-components or FGMs. The material chosen was pre-alloyed tungsten carbide alloy containing 12 % cobalt. It was mixed with different ratios of pure cobalt to give a series of powder blends that were then laser-sintered and formed a multi-layer specimen. The percentage of cobalt varied discretely from 12 % to 100 %. The results indicated balling in the cobalt-poor region and lack of complete melting in the cobalt-rich regions, which were more sensitive to scanning orientation than to scanning speed. Furthermore, microhardness measurements were taken at different locations of composition and the measured results were comparable to the reference values [16].

In research to generate a surface composite with a laser cladding technique, two kinds of AlSi-based powders were injected into the laser pool, and functionally gradient coatings (FGCs) were clad on cast Al alloy substrates in order to enhance their hardness and wear resistance (see Chapter 5) [17]. These graded microstructures formed during solidification led to a gradual variation of hardness in the cross-sections of these coatings. They indicated that the solidification behaviour of the laser pool played a key role in the formation of the functionally gradient coatings.

In an experiment with SiC coatings on aluminium produced by high-power lasers, the commonly encountered problems of the great difference of laser absorbance and thermal conductivity between the ceramic and aluminium were solved by heating the substrate to over 300 °C during laser processing [18]. Though an SiC–Al metal matrix composite was formed, due to the high temperature during the laser process some SiC particles were partially reacting with the liquid Al, resulting in the formation of a brittle $Al_4C_3$ phase. Moreover, $Al_4C_3$ is unstable in some environments such as water, methanol, HCl, etc., the composite possibly being susceptible to corrosive environments.

The sintering effect of CuSn as a pre-alloyed powder (bronze) and a mixed powder mixture was studied [19]. It was found that the melted tin failed to wet copper grains sufficiently in the instance of a powder mixture, while layer formation of pre-alloyed powder was successfully achieved at the same energy density.

Many laser–material processes are similar. Process variables in the literature are a good reference and starting point for further research, as summarised in Table 7.1.

**Table 7.1** List of process parameters from references

| Process | PROMET$^{TM}$ [20] | LENS$^{TM}$ [21] | Graded-boundary material [22] | Laser cladding [23] | Laser fabrication [24] |
|---|---|---|---|---|---|
| Laser type | $CO_2$ (10.6 μm) | Nd:YAG (1.06 μm) | $CO_2$ (10.6 μm) | — | — |
| Laser power (W) | 1700 | 200–300 | 1750 | 2000 | 1800 |
| Scanning speed (mm $s^{-1}$) | 25 | 6 to 9 | 8.33 | 4 | 13.3 |
| Preheat temperature (°C) | 250 | — | — | — | — |
| Layer thickness (mm) | 0.6–0.75 | 0.125–0.25 | 0.55 (sheet material) | — | — |
| Spot diameter (mm) | — | — | — | 5 | 15 (defocused) |
| Focal length (mm) | — | — | — | — | 250 |
| Bead width (mm) | 1.2–1.5 | — | — | — | — |
| Material | H10 tool steel powder | H13 tool steel powder SS316 (−150/+325 mesh) | Inconel 690 sheet and steel substrate | — | — |
| Shield gas | Argon | — | Argon | — | — |

### 7.3.3 Issues with Laser-Material Interactions

There are some issues in laser processing of powders that need to be considered. Firstly, the absorbance of two different materials might have a distinct difference. The absorbance of a material is defined as the ratio of the absorbed radiation to the incident radiation. It is usually obtained by measuring the reflectance (or reflectivity) instead. The absorbance plus reflectance of a material equals 1, providing that no transmission occurs.

The absorbance depends on the wavelength of incident radiation (e.g. laser), temperature, surface characteristics of the material, etc. The ambient gas, especially the presence of oxygen, is reported to also have effects on the absorbance [25]. The typical absorbance of metals at 10.6 μm is less than 0.1 (carbon steel of 0.03 and stainless steel of 0.09) and the absorbance of ceramics is generally much higher. On the contrary, the absorbance of metals and cermets are higher at wavelength 1.06 μm (Nd:YAG laser). For example, the absorbances of iron (Fe) and tungsten carbide (WC) powders were measured as 0.64 and 0.82 respectively [26].

Powders have higher absorbance than dense materials because the interstitial air (the air that occupies the space between powders) increases the absorption of the radiation by acting as a blackbody. Considering a powder mixture, a material may have a greater absorbance to the laser irradiation and would possibly melt before or at the same time as the other material begins to melt, even if the material has a higher melting temperature. This could cause an unwanted alloying effect or distorted structure.

Kim discussed some considerations in developing FGMs by laser processing [22]. In a metal/ceramic system, the great difference in laser beam absorbance could make processing both materials simultaneously difficult. Meanwhile, the rapid heating and cooling rates during the process could also cause cracks to form easily in the boundary region or in the ceramic part due to the large differences in thermal expansion coefficients and thermal conductivities. The solid solubility of both materials in each other may be extremely limited and make it difficult to obtain a graded structure by laser processing. The development of a graded boundary material of Inconel690 and steel was achieved by layering sheet material sequentially and applying laser surface alloying treatment up to four times, where the major processing parameters were laser power of 1750 W, scanning speed of 0.5 m $min^{-1}$ and 50 % overlapped pass of scanning [22].

The materials implications can be severe if the laser completely melts the powder in direct SLS or similar processes. In addition to significant residual stresses caused by extremely high local heating, the control on the microstructures is also critical. Steel tool properties depend on the austenite/ferrite solid-state phase transformation, whereas control of a part's microstructure produced direct metal SLS is still not yet routinely achievable [27].

## 7.4 Modelling and Software Issues

### *7.4.1 Compositional Profile*

Unlike homogeneous materials, the properties of FGMs not only vary with their composition but also depend on the connectivity of the network structure. Figure 7.5 shows a schematic representation of the common microstructure occurring in a material as the content of the second phase is increased. At low volume fractions the second phase exists as isolated particles dispersed within a matrix (a). As the content of the second phase increases, the particles begin to have contact and form agglomerated clusters (b). As it increases further, a critical microstructural transition takes place, where the second phase is no longer dispersed but rather becomes interconnected over long distances (c to e). The transition has a profound effect on the material properties, e.g. thermal or electrical conductivity, and a small change of composition will therefore result in a distinct variation in properties [28].

Connectivity in the microstructure can be studied using percolation models. These models treat the microstructure as a network of nodes and interconnections in which composition is measured by the number of completed connections or occupied nodes. The percolation problem can then be analysed using fractal theory. The phase connectivity in a two-phase mixture is described by the Betti number, which is a topographic parameter to characterise this unique feature [7]. The critical composition for second phase connectivity has been found to depend on several factors, including the packing arrangement, particle shape of the second phase and the size ratio of the constituents [7].

Modelling an FGM's topology and composition was approached by Jackson and co-workers. Their method was based on subdividing the solid model into subregions and associating analytic composition blending functions with each region. These blending functions defined the composition throughout the model to achieve local composition control [29].

There are many possibilities to design and model the compositional change. For a two-phase system, the phase distribution is commonly described by

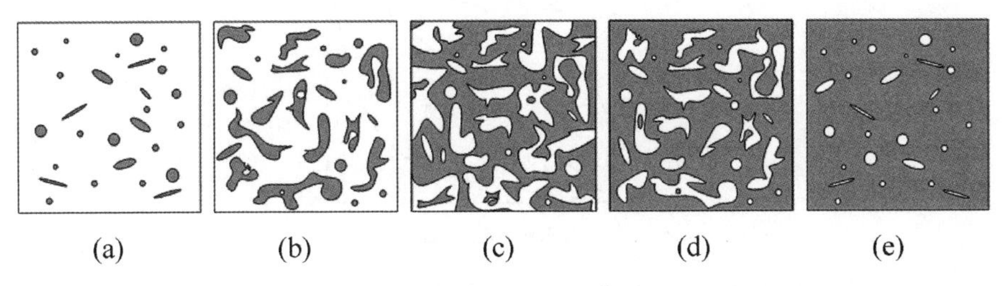


**Figure 7.5** Schematic of microstructural changes in cross-sections of an FGM

a power law equation:

$$x_B = \left(\frac{d}{L}\right)^n \tag{7.1}$$

and

$$x_A = 1 - x_B$$

where $x_A$ and $x_B$ represent the volume fraction of material (phase) A and B respectively, $d$ is the distance into the graded structure, $L$ represents the total length of the FGM and $n$ is the compositional exponent, as shown in Figure 7.6.

### *7.4.2 Software Issues*

Solid freeform fabrication (SFF) processes are characterized by their high degree of automation. From computer aided design (CAD) data, parts must be made without much effort when programming the manufacturing steps. To be able to produce FGM parts by these technologies it is necessary to receive and process data that carry the information about the gradients. One reason for the great success of SFF was the use of a common and easy way to transfer data from CAD systems to the computer aided manufacture (CAM) systems of the different SFF technologies. The STL (Standard Triangulation Language) format was accepted as a standard for the industry. In the same way, a new format will be necessary and research has started to appear in

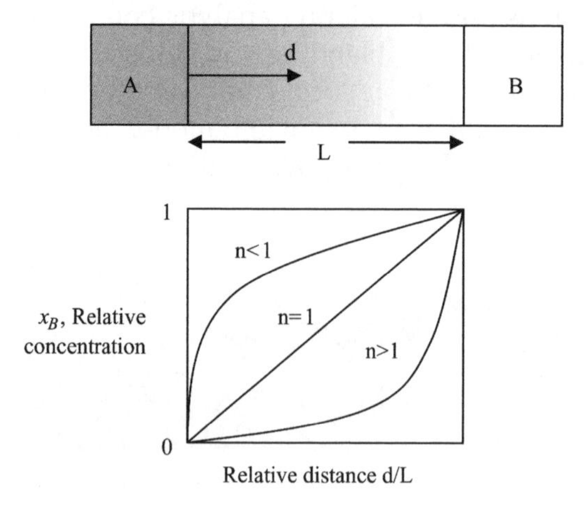


**Figure 7.6** Different compositional profiles obtained by varying the compositional exponent in the power law equation

this direction for FGMs. Currently, researchers have implemented the representation of heterogeneous objects in electronic data. Rvachev *et al.* [30] showed the possibility of using the *R*-function method to represent FGM solids. These *R*-functions were used to make the gradient from the boundaries to the interior of the solid. Another approach was presented by Jackson [31]. His alternative was to design the gradients in the parts by creating a tetrahedral mesh where the gradients could be assigned to the mesh elements. In the same way, the Voxel-based representation uses a three-dimensional spatial grid to define finite material properties to each cell of the grid. Knoppers *et al.* showed an STL/Voxel hybrid modeller for designing functionally graded parts [32]. The software could create gradients based on user inputs. The advantage of this software was that it could directly export slices of the model to any three-dimensional printing machine. With some processing it can be used to build a three-dimensional part [32]. Nevertheless, Voxel representation requires great computer power and processing. Kumar and Wood [33] have shown the possibility of using an $r_{\mathrm{m}}$-object model to design FGM solids. It also represents the heterogeneity of the solid by associating the material distribution with functions. Patil *et al.* [34] suggested an informtion model to represent heterogeneous objects in Layered Manufacturing using information modeling methodology developed for ISO 10303 (informally known as STEP – Standard for the Exchange of Products model data). It has been discussed that the $r_m$ *object model* (a set based approach) could represent multi material objects. Bhashyam, Shin and Dutta [35] presented a CAD system for generating FGM parts. In this system, the $r_{\mathrm{m}}$-object was also used for designing heterogeneous CAD solid models. Some advances on designing FGM parts have been made already on commercial CAD packages. Liu *et al.* [36] and Siu and Tan [37] have developed applications for popular CAD software where it is possible to design solid parts and to apply graded material compositions. Liu *et al.* [36] developed the application for the commercial CAD package Solidworks and the approach was to produce the gradient based on the distance from features of the model. Siu and Tan [37] implemented the concept of a grading source for designing the gradient in the CAD model on the Unigraphics software.

## 7.5 Characterisation of Properties

The simplest method to estimate the transitional composition in FGMs is to apply the rule of mixtures. Using a similar method to calculate a classic problem of heat conduction in a body consisting of multiple layers, the thermal conductivity and temperature gradient of a one-dimensional FGM can then be determined.

### 7.5.1 Thermal Properties

#### Coefficient of Thermal Expansion

The thermal expansion or contraction of a material is given by the coefficient of thermal expansion and the temperature difference experienced. Due to the varied composition and structure in an FGM, thermal expansion from different components significantly affects the FGM's function and mechanical performance. Research has been devoted to studying the thermal expansion in FGMs in order to optimise the graded composition through the relaxation of thermal stress. The coefficient of thermal expansion (CTE) can be measured with a dilatometer. A detailed measurement in two or three dimensions can be done using laser interferometry or digital image correlation [7].

#### Thermal Conductivity

The heat flux $\dot{q}$ transferred into a medium can be written as

$$\dot{q} = -\lambda \frac{\mathrm{d}T}{\mathrm{d}x} \tag{7.2}$$

where $\lambda$ represents the material's thermal conductivity, $T$ the temperature and $x$ distance respectively. If the heat flux flows through the direction of change in composition, the effective thermal conductivity $\lambda_e$ of the graded material can be written as adding the thermal resistance together in equation (7.2):

$$\frac{1}{\lambda_e} = \frac{1}{L} \int_0^L \frac{\mathrm{d}x}{\lambda(x)} \tag{7.3}$$

where $L$ is the total length of the FGM. Nevertheless, the thermal conductivity $\lambda(x)$ as a function of location $x$ is usually unknown or hard to obtain. An approximation will be to treat the graded material as a laminate of numerous layer thickness $l_i$:

$$\begin{aligned} \frac{1}{\lambda_e} &= \frac{1}{L} \sum_{i=1}^{n} \frac{l_i}{\lambda_i} \\ L &= \sum_{i=1}^{n} l_i \end{aligned} \tag{7.4}$$

For simplicity, the layer thickness is assumed to be uniform. The composition of phase B at the $i$th layer is, according to the equation (7.1),

$$x_i = \left(\frac{il}{L}\right)^n \tag{7.5}$$

wherein the index $i$ starts from zero. The thermal conductivity of the $i$th layer follows the rule of mixture as

$$\lambda_i = \lambda_A(1 - x_i) + \lambda_B x_i \tag{7.6}$$

By inserting equation (7.6) into equation (7.4), the effective thermal conductivity of the laminate can be determined as

$$\lambda_e = \left(\frac{1}{L}\sum_{i=1}^{n}\frac{l_i}{[\lambda_A(1 - x_i) + \lambda_B x_i]}\right)^{-1} \tag{7.7}$$

### Temperature Difference across a Laminate under a Thermal Load

If the graded structure is assumed to be approximated by a laminated structure of uniform layer thickness, the effective thermal conductivity can be calculated from equations (7.4) and (7.6). The temperature across the laminate material with length $L$ can be obtained from the following equation:

$$\Delta T = \frac{\dot{q}L}{\lambda_e} \tag{7.8}$$

Figure 7.7 shows the temperature difference as a function of the compositional exponent $n$ under a given thermal load $\dot{q}$. In practice, this means that

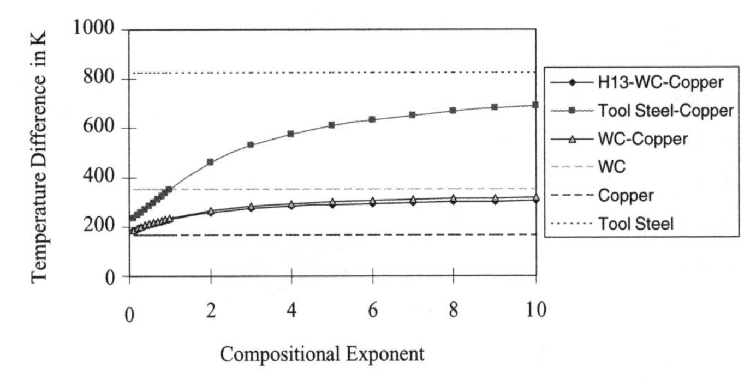


**Figure 7.7** Calculated temperature difference in a laminate under a given thermal condition

the temperature gradient can be controlled by optimising the compositional exponent $n$. In this example, the temperature difference across the laminate increases with increasing $n$. A great temperature gradient in a material also implies increasing thermal stresses which may have adverse effects on the material properties.

### Design of the Compositional Profile for a Laminate under a Certain Thermal Boundary Condition

Another common situation is to find out how to design the graded composition of a material that is forced to sustain a temperature gradient. For example, when the space shuttle enters the atmosphere, the aircraft will experience a huge temperature gradient across the materials. Similarly, thermal barrier materials need to perform under a huge temperature difference across their structure. In a tooling application, a casting die also experiences a great temperature difference existing from its surface to its core.

In Figure 7.8, a laminate of varying composition of materials A and B is put under a given thermal load $\dot{q} = 3 \times 10^6$ W m$^{-2}$. Supposing the temperature difference across the laminate to be 1000 K, the necessary thickness of graded structure for a specified compositional exponent can be looked up from the curves and vice versa, which is helpful in designing and optimising an FGM's thermal properties.

### *7.5.2 Mechanical Properties*

The microstructure of an FGM changes spatially. Therefore, the toughness and strength of an FGM cannot be evaluated simply by conventional fracture mechanics and standard mechanical testing [7]. The analysis and measurement of elastic properties and the influence of thermal stresses will be addressed in the following text.

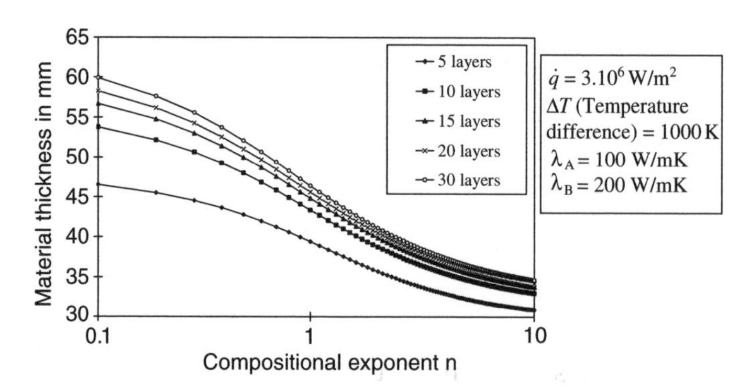


**Figure 7.8** The relationship of the material length of graded composition and the compositional exponent $n$ for a given thermal condition

### Elastic Modulus

For determining the overall Young's modulus of an FGM, the usual stress–strain curve can be measured by attaching a strain gauge to the sample's surface during a four-point bending test. The dependence of Young's modulus on an FGM's composition can be estimated by measuring the flexural resonant frequencies of a rectangular bar specimen using a forced-resonance technique.

Alternatively, the distribution of the Young's modulus and the Poisson's ratio can be measured simultaneously by using a line-focus-beam (LFB) acoustic microscope [7]. The relative intensity of the received surface-reflected wave is measured as a function of the distance between the acoustic lens and the specimen. From the measured surface acoustic waves, the Young's modulus and the Poisson's ratio can be derived [7].

### Deformation and Strength

Both the strength and the deformation of an FGM are commonly evaluated by tensile tests and three- or four-point bending tests. The residual stress, originating from cooling of the processing temperature, has a strong influence on the tensile behaviour of FGMs by shifting the stress state to tension or compression, and by strain hardening when the elastic limit is exceeded [7].

### Thermal Stresses

Thermal stresses come from the constraints in deformation of components (avoid high stress at cracks, etc.) that are subject to certain thermomechanical boundary conditions such as inhomogeneous temperature distribution in a homogeneous component, uniform heating of an heterogeneous component (e.g. a metal/ceramic joint) or rapid heating and cooling of a part's surface. For FGMs, these conditions are (often) all involved.

Therefore, to develop an FGM to be and in a thermally fluctuating environment, one must know the thermal stress present in each of these conditions. The relation between the estimated thermal stress and the observed fracture mode can be used in defining the design criteria [7].

## 7.6 Deposition Systems

### *7.6.1 Local Composition Control*

Despite the possibilities of manufacturing an FGM by RM techniques all future success depends on the development of local composition control (LCC). To give the freedom of materials and design, layers must consist of

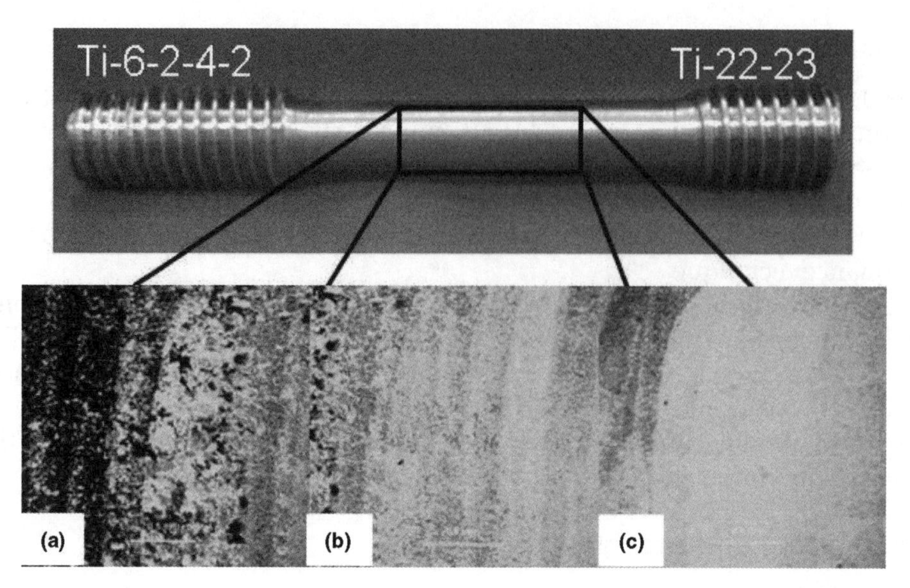


**Figure 7.9** Metallographic cross-sections from the graded transition region: (a) close to gamma titanium aluminide end, (b) near the centre of the gradient region and (c) from the orthorhombic end. Reproduced with permission of Optomec. Inc

multi-materials. This means that the layer deposition system must have the ability to deposit more than one material, in different proportions and in specific regions, with accuracy. Laser cladding technologies have the advantage of not pre-placing the layer as the powder is coaxially sprayed in the laser focus. Ensz, Griffith and Reckaway [38] produced FGM thin walls of H13 and M300 steels controlling the powder composition of the spray. By this technology (LENS$^{TM}$, Laser Engineering Net-Shape), Chavez also presented gradients of titanium alloys [39] (see Chapter 5). In Figure 7.9, transitions between gamma titanium aluminide (Ti–48Al–2Nb–2Cr) and orthorhombic titanium aluminide (Ti–22Al–23Nb) are shown.

For pre-placed powder beds, Kumar *et al.* [40] had investigated the use of fine hoppers to deposit powders. The hoppers used were attached to a numeric controlled table and the powder deposition was tested by gravity, gas pressure assisted and vibration. This system could be used to deposit different materials using as many hoppers as necessary depending on the graded transition needed. The apparatus could be similar to an inkjet printer but instead of ink it would deposit powders over the bed. Yang and Evans [41] had studied the same principle of hoppers but with capillary diameters. The powder deposition was excited by acoustic vibration. A more complex process for local composition control is by Shape Deposition Manufacturing (SDM) where FGM parts could be produced by depositing multi materials using powder feeders and subsequently melted by a laser. By this technique it is possible to produce embedded flexible materials in multi-material prototypes (by using a hybrid of photolithography and SDM) [42] and graded metallic parts [43].

## 7.7 Applications

### 7.7.1 Aerospace

The aerospace sector is one of the sectors that shows more interest on the development of FGMs. NASA and JAXA had shown high interest in the development of this technology. As the cost for launching space shuttles depends on the weight they lift into space, creating parts with less connection elements is a solution. Also, the optimum design, balancing weight and performance can reduce costs for launching. Performance and safety are always of great concern in aerospace. The heat shield protection of space shuttles could be optimised to reduce weight and increase reliability by using FGM plates.

Bhatt [44] showed the use of silicon nitride ceramics for turbine nozzle vanes. Silicon nitride supports high temperatures but has a high processing cost and low impact resistance. Grading using SFF technologies, powder metallurgy and gel casting were studied as alternatives to produce viable parts. Aboudi, Pindera and Arnold [45] presented the use of FGM for tailoring the response of structural components by high-order theory. This was intended to analyse the performance of the gradient by the microstructure modelling using finite element model (FEM) analysis. Later work presented by Arnold [46] applies the theoretical model for simulating the response of FGM for cooling thermal shield plates for space shuttles.

### 7.7.2 Sporting Goods

Various sports components and accessories can benefit from the use of FGMs. For example, WC carbide tips were manufactured for baseball shoes in order to reduce the wear. Between the WC tip and the structure of the shoes a graded material region was produced to reduce the impact on the tips [47]. This is an excellent example of how FGMs could be used to produce high-performance sporting goods. With the same idea of wear reduction, ice climbing tools such as crampons and ice axes could be graded with tungsten carbide in the sharp edges that picks the ice. Rock climbing technical kits also could have the same advantage to reduce the wear such as on cam surfaces.

### 7.7.3 Medical

Among the applications for FGMs, medical implants appear a strong field of research (see Chapter 11). The aim of using FGMs to produce implants is to use the mechanical properties of the alloys, ceramics and composites grading with biocompatible materials. Thus a bone implant can be formed by a strong and tough material in the core with graded bone tissue compatible material to the surface and low friction material in the joints.

Watari *et al.* [48] presented a method for the fabrication of dental implants of titanium and ceramic hydroxyapatite. The base of the implant was made of hydroxyapatite to be inserted in the jaw. The implant was titanium graded to the top in order to receive the teeth prosthesis. The experiment showed successful bone tissue growing around the insert. The implant was produced by grading the titanium and hydroxyapatite powders inside a silicon rubber mould that was subjected to cold isostatic pressure and then sintered in a furnace. Chu *et al.* [49] also studied FGMs with titanium and hydroxyapatite and the results showed that better mechanical properties for the implant were obtained as no interfacial problem was detected. Roop Kumar and Maruno [50] also added glass to hydroxyapatite powder and coated titanium implants for better biocompatibility. Glass and glass-matrix/zirconia graded layers were used to coat alumina substrates by Vitale Brovarone *et al.* [51]. The grading was necessary due to the different expansion coefficient mismatch between the substrate and the coatings. The coatings provided good biocompatibility and the alumina provided good mechanical properties. Castillo *et al.* [52] presented the fabrication of biocompatible FGM by combustion synthesis. Results have been found to be better when producing gradients in low gravity to form a more homogeneous and continuous smooth gradient. Also, pores and grains formed in low gravity were rounded differently from those formed under terrestrial experiments with lamellar and longitudinal structures. The structural format of the bone implant can be affected by these differences in the formed structure. It was shown that materials produced in low gravity can increase the rate of tissue growing around the implant. Khor *et al.* [53] produced FGM structures of hydroxyapatite coated with Ti–6Al–4V by plasma spray. Different percentages of Ti–6Al–4V were added to the hydroxyapatite for graded layers. The results indicated good cohesion between the layers without a strong distinction between the layers. The Ti–6Al–4V provides good mechanical strength and the hydroxyapatite provides bone biocompatibility. Pompea *et al.* [54] describe many cases of implants based on functionally graded materials. Knee joint implants could be manufactured with ultra-high molecular weight polyethylene fibre reinforced high-density polyethylene in gradients for producing low friction joints, but strengthened close to the bone. Also, hydroxyapatite could be produced in different ways and porosity could be controlled to make bone tissue regeneration more efficient.

## Acknowledgement

The authors are gratefully acknowledges Wei-Nien Su for his contributions and EPSRC for the support of this work.

## References

1. Hirai, T. (1996) Functional gradient materials, in *Materials Science and Technology – A Comprehensive Treatment*, (eds R.W. Cahn *et al.*), Vol. 17B, *Processing of Ceramics*, Part II, VCH Verlagsgesells chaft, Weinheim and Cambridge, pp. 293–363.
2. POM (2000) *Direct Metal Deposition (DMD)*, POM Group, Inc., http://www.pom.net (Accessed 12 July 2002).
3. Bever, M.B. and Duwez, P.F. (1972) *Material Science Engineering*, **10**, 1–8.
4. Shen M. and Bever M.B. (1972) *Journal of Material Science*, **7**, 741–6.
5. Mortensen, A. and Suresh S. (1995) *International Material Review*, **40**, 239–65.
6. Neubrand, A. and Rodel, J. (1997) *Zeitschrift Metallkd*, **88**, 358–71.
7. Miyamoto, Y., Kaysser, W.A., Rabin, B.H., Kawasaki, A. and Ford, R.G. (1999) *Functionally Graded Materials*, Kluwer Academic Publishers, Boston.
8. Hirai, T. (1996) In R.J. Brook (Ed.), *Materials Science and Technology*, Vol. 17B, *Processing of Ceramics*, Part 2, VCH Verlagsgesellschaft, Weinheim and Cambridge, pp. 292–341.
9. Suresh, S. and Mortensen, A. (1998) *Fundamentals of Functionally Graded* Materials, Institute of Materials, London.
10. German, R.M. (1996) *Sintering Theory and Practice*, John Wiley & Sons, Inc., New York.
11. Shen, J., Keller, B. and Stierlen, P. (1997) Rapid Prototyping – Material- und Prozessentwicklung fuer die Lasersinter-Technologie, in 15th Stuttgarter Kunststoff-Kolloquium, Stuttgart, IKK and IKP, Universitaet Stuttgart, pp. 10.1–10.4.
12. Van der Schueren, B. and Kruth, J.P. (1995) Powder deposition in selective metal powder sintering, *Rapid Prototyping Journal*, **1**(3), 23–31.
13. Kruth, J.P., Leu, K.C. and Nakagawa, T. (1998) Progress in additive manufacturing and rapid prototyping, *Annals of the CIRP*, **47**(2), 525–40.
14. EOS GmbH (2002), *Eosint M250x*, EOS GmbH, Germany, http://www.eos-gmbh.de/pag/file/helpfiles/020300_info_m250.pdf (Accessed 18 April 2002).
15. Schwendner, A., Banerjee, R., Collins, P.C., Brice, C.A. and Fraser, H.L. (2001) Direct laser deposition of alloys from elemental powder blends, *Scripta Materialia* **45**, 1123–9.

16. Jepson, L., Perez, J., Beaman, J., Bourell, D. and Wood, K. (1999) Initial development of a multi-material selective laser sintering process (M2SLS), in 8th European Conference on *Rapid Prototyping and Manufacturing*, Nottingham, 6–8 July 1999, University of Nottingham, pp. 367–84.
17. Pei, Y.T. and De Hosson, J.T.M. (2000) Producing functionally graded coatings by laser-powder cladding, *Journal of Metals, Part e*, **52**(1), http://www.tms.org/pubs/journals/JOM/0001/Pei/Pei-0001.html (web-edition).
18. Vreeling, J.A., Ocelik, V., Pei, Y.T. and De Hosson, J.T.M. (1999) *Microstructure Study of Ceramic Coatings on Aluminium Produced by High Power Lasers*, Rijkuniversiteit Groningen, http://www.phys.rug.nl/mk/research/98/interfaces_lasers.html (Accessed 23 March 2002).
19. Klocke, F., Celiker, T. and Song, Y.-A. (1995) Rapid metal tooling, *Rapid Prototyping Journal*, **1**(3), 32–42.
20. Peterseim, J. and Luck, J.M. (1997) Direct metal prototyping by the laser fusion of metal powder, in International Conference on *Competitive Advantages by Near-Net-Shape Manufacturing (NNS'97)*, Bremen, Germany, 14–16 April 1997, DGM Informationsgesellschaft mbH, pp. 319–26.
21. Griffith, M.L. and Schlienger, M.E. (1998) Thermal behavior in the LENS process, in 9th Solid Freeform Fabrication Symposium, Austin, Texas, 10–12 August 1998, University of Texas at Austin, pp. 89–96.
22. Kim, T.H. (1998) Development of graded-boundary material by laser beam, *Journal of Laser Applications*, **10**(5), 191–8.
23. Wang, P.-Z., Qu, J.-X. and Shao, H.-S. (1996) Cemented carbide reinforced nickel-based alloy coating by laser cladding and the wear characteristics, *Materials and Design*, **17**(5/6), 289–96.
24. Kathuria, Y.P. (1997) Laser microfabrication of metallic parts, *Proceedings of the SPIE, The International Society of Optical Engineering*, **3102**, 112–19.
25. Tolochko, N.K., Laoui, T., Khlopkov, Y.V., Mozzharov, S.E., Titov, V.I. and Ignatiev, M.B. (2000) Absorptance of powder materials suitable for laser sintering, *Rapid Prototyping Journal*, **6**(3), 155–60.
26. Xie, J., Kar, A., Rothenflue, J.A. and Latham, W.P. (1997) Temperature-dependent absorptivity and cutting capability of $CO_2$, Nd:YAG and chemical oxygen–iodine lasers, *Journal of Laser Applications*, **9**, 77–85.
27. Derby, B. (1999) Materials issues in rapid manufacture of metal and ceramic parts, in Time-Compression Technologies '99 Conference, East Midlands Conference Centre, Nottingham, 12–13 October 1999, Rapid News Publications plc, pp. 133–9.
28. Su, W.-N. (2002) *Layered Fabrication of Tool Steel and Functionally Graded Materials with a Nd:YAG Pulsed Laser*, Loughborough University, Loughborough.
29. Jackson, T.R., Liu, H., Partikalakis, N.M., Sachs, E.M. and Cima, M.J. (1999) Modeling and designing functionally graded material components

for fabrication with local composition control, *Materials and Design*, **20**, 63–75.
30. Rvachev, V.L., Sheiko, T.I., Shapiro, V., Tsukanov, I. (2001) Transfinite interpolation over implicitly defined sets, *Computer-Aided Geometric Design*, **18**, 195–220.
31. Jackson, T.R. (2000) Analysis of functionally graded material object representation methods. Submitted thesis to the Departamento of Ocean Engineering, Massachusetts Institute of Technology, Cambridge, Massachusetts, June 2000.
32. Knoppers, G.E., Gunnink, J.W., van den Hout, J. and van Vliet, W.P. (2004) The reality of functionally graded products, SFF2004, pp. 38–43.
33. Kumar, A. and Wood, A. (2004) Representation and design of heterogeneous components, in Proceedings of the 1999 Solid Freeform Fabrication Symposium, Austin, Texas, pp. 179–86.
34. Patil, L., Dutta, D., Bhatt, A.D., Jurrens, K., Lyons, K., Pratt, M.J. and Sriram, R.D. (2002) A proposed standards-based approach for representing heterogeneous objects for layered manufacturing, *Rapid Prototyping Journal*, **8**(3), 134–46, MCB University Press.
35. Bhashyam, S., Shin, K.H. and Dutta, D. (2000) An integrated CAD system for design of heterogeneous objects, *Rapid Prototyping Journal*, **6**(2), 119–35, Emerald Group Publishing Limited.
36. Liu, H., Maekawa, T., Patrikalakis, N.M., Sachs, E.M. and Cho, W. Methods for feature-based design of heterogeneous solids, *Computer-Aided Design*, **36**, 1141–59, Elsevier Ltd.
37. Siu, Y.K. and Tan, S.T. (2002) Representation and CAD modeling of heterogeneous objects, *Rapid Prototyping Journal*, **8**(2), 70–75.
38. Ensz, M.T., Griffith, M.L. and Reckaway, D.E. (2002) Critical issues for functionally graded material deposition by laser engineered net shaping (LENS™), in Proceedings of the 2002 MPIF International Conference on *Metal Powder Deposition for Rapid Manufacturing*, San Antonio, Texas.
39. Chavez, P. (2000) LENS fuels paradigm shift in modern manufacturing, from the Inside Out, Metal Powder Report, September 2001 Issue, http://www.metal-powder.net/septfeat2.html (Accessed on 20 October 2004).
40. Kumar, P., Santosa, J.K., Beck, E. and Das, S. (2004) Direct-write deposition of fine powders through miniature hopper-nozzles for multi-material solid freeform fabrication, *Rapid Prototyping Journal*, **10**(1), 14–23.
41. Yang, S. and Evans, J.R.G. (2004) Acoustic control of powder dispensing in open tubes, *Powder Technology*, **139**, 55–60.
42. Hatanaka, M. and Cutkosky, M. (2003) Process planning for embedding flexible materials in multi-material prototypes, in Proceedings of DETC'03: 2003 ASME Design Engineering Technical Conferences and

Computers and Information in Engineering Conference, Chicago, Illinois, 2–6 September 2003.
43. Fessler, J.R., Nickel, A.H., Link, G., Prinz, EB. and Fussell, P. (1997) Functional gradient metallic prototypes through shape deposition manufacturing, in Proceedings of the 1997 Solid Freeform Fabrication Symposium, Austin, Texas, 11–13 August 1997, pp. 521–528.
44. Bhatt, R.T. (2000) Feasibility of actively cooled silicon nitride airfoil for turbine applications demonstrated, in *Research and Technology 2000*, Glenn Research Center at Lewis Field, Cleveland, Ohio, p. 34.
45. Aboudi, J., Pindera, M.J. and Arnold, S.M. (2000) Higher-order theory for functionally graded materials, in *Research and Technology 2000*, Glenn Research Center at Lewis Field, Cleveland, Ohio, pp. 105–6.
46. Arnold, S.M. (2001) Higher-order theory – structural/microanalysis code (HOT–SMAC) developed, in *Research and Technology 2001*, Glenn Research Center at Lewis Field, Cleveland, Ohio, pp. 128–9.
47. JAXA and JST (2004) FGMs database: introduction of applications, http://fgmdb.nal.go.jp/e_index.html (Accessed August 2004).
48. Watari, F., Yokoyama, A., Saso, F., Uo, M. and Kawasaki, T. (1997) Fabrication and properties of functionally graded dental implant, *Composites Part B*, 5–11, Elsevier Science Limited.
49. Chu, C., Zhu, J., Yin, Z. and Wang, S. (1999) Hydroxyapatite–Ti functionally graded biomaterial fabricated by powder metallurgy, *Materials Science and Engineering*, **A217**, 95–100, Elsevier Science Limited.
50. Roop Kumar, R. and Marino, S. (2002) Functionally graded coatings of HA–G–Ti composites and their *in vivo* studies, *Materials Science and Engineering*, **A334**, 156–62, Elsevier Science Limited.
51. Vitale Brovarone, C., Verne, E., Krajewski, A. and Ravaglioli, A. (2001) Graded coatings on ceramic substrates for biomedical applications, *Journal of the European Ceramic Society*, **21**, 2855–62, Elsevier Science Limited.
52. Castillo, M., Moore, J.J., Schowengerdt, F.D., Ayers, R.A., Zhang, X., Umakoshi, M., Yi, H.C. and Guigne, J.Y. (2003) Effects of gravity on combustion synthesis of functionally graded biomaterials, *Advances in Space Research*, **32**, 265–70, Elsevier Science Limited.
53. Khor, K.A., Gu, Y.W., Quek, C.H. and Cheang, P. Plasma (2003) spraying of functionally graded hydroxyapatite–Ti–6Al–4V coatings, *Surface and Coatings Technology*, **168**, 195–201, Elsevier Science Limited.
54. Pompea, W., Worch, H., Epple, M., Friess, W., Gelinsky, M., Greil, P., Hempele, U., Scharnweber, D. and Schulte, K. (2003) Functionally graded materials for biomedical applications, *Materials Science and Engineering*, **A362**, 40–60, Elsevier Science Limited.

# 8

# Materials and Process Control for Rapid Manufacture

Tim Gornet
*The University of Louisville*

## 8.1 Introduction

Since the advent of the first Rapid Prototyping (RP) machines in the mid 1980s people have futuristically predicted the evolution of the technology into the creation of a real 'Replicator' of the Star Trek TV show and book series fame. This standard piece of starship equipment could immediately copy and provide almost any meal or equipment desired by the crew. While still quantum leaps from the machine portrayed on TV, the RP world is beginning to see applications of end-use parts directly manufactured on RP equipment incorporated in everything from aerospace to military to medical fields.

The Boeing Company has flight certified a nylon material that is currently in use for non-structural parts on F18 aircraft (see Chapters 15 and 16). Siemens and Phonak, hearing aid manufacturers, create hearing aid shells directly out of nylon with selective laser sintering (SLS) (see Chapter 12). Widex is another hearing aid manufacturer that is creating shells from a custom stereolithography (SL) resin. Invisalign uses stereolithography in the manufacturing stage of their invisible braces. Rapid Manufacturing (RM) for low-volume, high-value parts and mass customization lead the incorporation of these technologies into their design and manufacture cycles.

*Rapid Manufacturing: An Industrial Revolution for the Digital Age*
Editors N. Hopkinson, R.J.M. Hague and P.M. Dickens 

The currently available equipment is able to create parts that are unable to be manufactured conventionally by taking advantage of the layerwise additive manufacturing process. Complex geometries with undercuts and channels can be fabricated in a single part that would normally require multiple pieces and processes to achieve a similar result. These parts can be made out of common engineering thermoplastics such as polyamides, acrylonitrate butadiene styrene (ABS), polycarbonate, polyphenylsulfone (PPSF) to metal parts from titanium, stainless steel and gradient metal structures (see Chapter 7).

It is easy to envision the advantages of solid freeform fabrication (SFF) on removing manufacturing constraints on geometric shapes that can be built. It is more difficult to actually put the process to the test and create a functional, end-use part due to the limited material and process selections available. The body of knowledge for the true mechanical properties of RP materials is quite small. Properties for all build orientations are not readily available from material or machine manufacturers. Material vendors rarely list the machine settings, build styles, or orientations of the ASTM (American Society for Testing and Materials) bars for the data that are available. Properties above or below room temperature and over extended timeframes are rarer and not publicly documented. This limits the design engineer from creating true structurally loaded components unless a large volume of mechanical testing is completed. This leaves the majority of the testing of mechanical properties to the end user.

The increased utilization of additive manufacturing for end-use parts is a materials game. There must be a material available to fulfill the design requirements of the desired part in terms of mechanical properties, physical properties, resistance to corrosion or chemicals, operating temperatures, flame retardancy, etc. Currently there are limited materials for each additive fabrication technology as compared to the myriad of metals and plastics for conventional manufacturing.

In this chapter, the most likely technologies will be discussed separately as to their feasibility for RM, detailing the materials currently offered and what properties are known for those materials. In addition, the development and application possibilities for increasing the material options will be discussed.

## 8.2 Stereolithography

Stereolithography (SL), developed by Chuck Hull and introduced by 3D Systems, Valencia, California, in 1988 has the largest number of materials available for a process. The relatively large number of machines in the field and multiple vendors of photopolymer resins have fueled this growth. There are over 40 resin formulations available today. New applications of the technology include printer head based SL.

Early photopolymer materials for SL were strong but quite brittle. Today there are many resins that fulfill specific market segments that require specific mechanical or material properties. There are flexible materials, resins that mimic the mechanical properties of common thermoplastics, nano and ceramic filled resins, and specialty resins for use in the medical arena. Widex (Denmark) is creating hearing aid shells directly with a photopolymer resin. This material is approved for long-term skin contact and has passed cytotoxicity testing. Invisalign (USA) utilizes multiple SLA-7000 systems for creating individual vacuum form molds via stereolithography apparatus (SLA) for their invisible braces system.

Previous work by Mueller (2004) provides an excellent overview of the various mechanical properties available in the SL resins and some comparisons to fused deposition modeling (FDM) and selective laser sintering (SLS). The real issue is matching the required properties for the application of the part to an available material. It is important to understand that one is not going to 'replace' polypropylene with SOMOS 11120 resin. It has some similar properties, but falls short in others. The choice has to be made on which properties are needed and which are secondary.

The broad range of flexural modulii available from rapid prototyping material systems is shown in Figure 8.1. The new filled material systems available that incorporate traditional or nanocomposite technologies

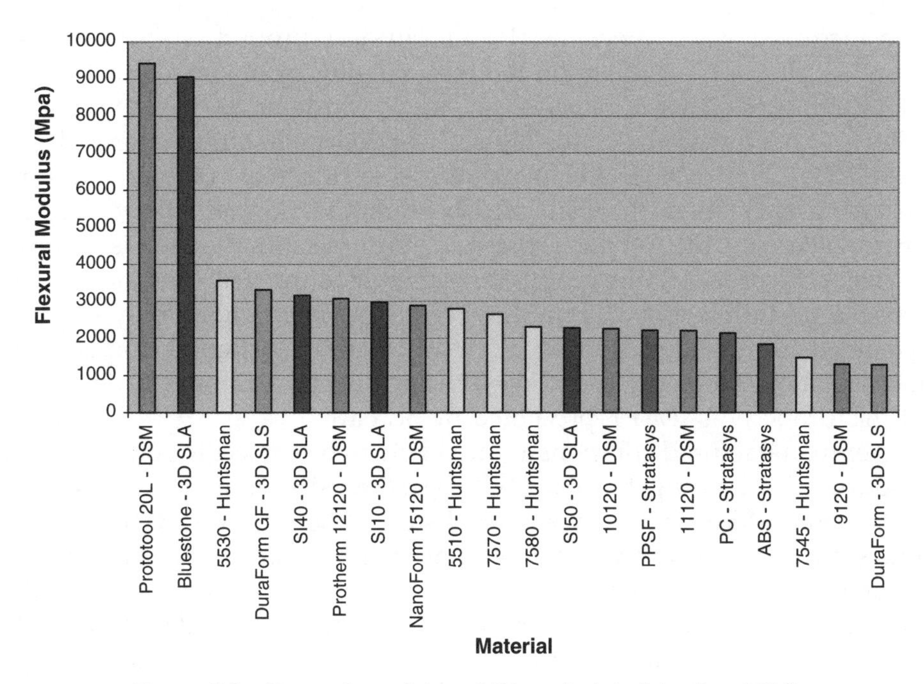


**Figure 8.1** Flexural modulus of RP materials (Mueller, 2004)

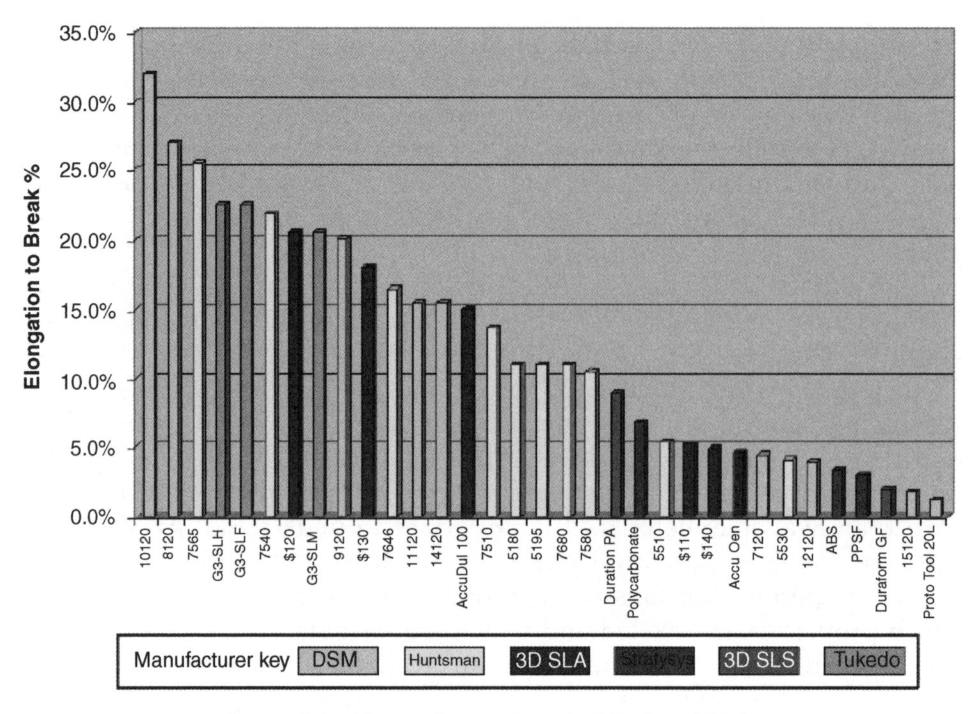


**Figure 8.2** Elongation to break (Mueller, 2004)

have increased some mechanical and physical properties tremendously. Figure 8.2 shows the elongation to break ranging from just about 1 to over 33 %. Engineering thermoplastics can run upwards of 200 % elongation.

The new filled materials have higher heat deflection temperatures (HDTs) as well (see Figure 8.3). Achieving these results may require additional processing such as a thermal cure beyond the normal ultraviolet (UV) curing. Where the SL systems represent thermoplastics the least is in impact resistance. Figure 8.4 is the graph of notched Izod impact resistance. Almost all resins are well below 0.5 J $cm^{-1}$, whereas impact modified polypropylene in molded form achieves about a 5 J $cm^{-1}$ reading. Granted, impact resistance is not only material but also geometry dependent in the designed part but this property is least represented by SL parts.

Layer manufactured components are likely to be subject to some degree of anisotropy. To investigate this, ASTM D638 tensile bars were provided by Carl Dekker, Met-l-Flo, an experienced stereolithography service bureau in Illinois, USA, out of several different materials built at standard parameters. Multiple parts oriented in the various axes were provided for evaluation. Bars were tested within two weeks of being built. The results of the ASTM D638 mechanical testing and the data provided by DSM SOMOS product data sheets are shown in Tables 8.1 and 8.2 respectively. It should be noted

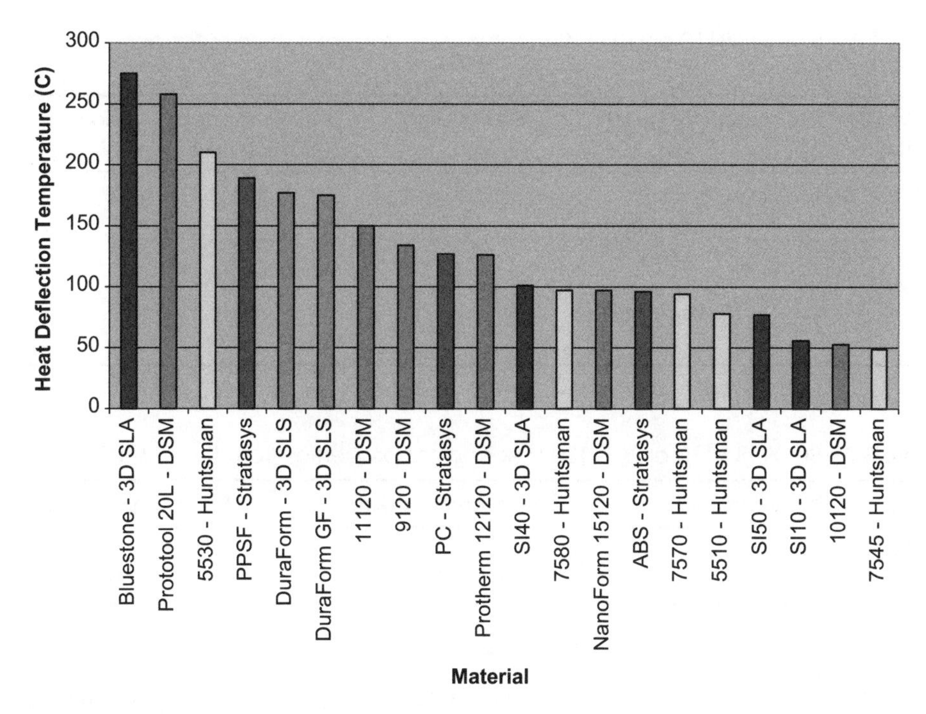


**Figure 8.3** Heat deflection temperatures of RP materials (Mueller, 2004)

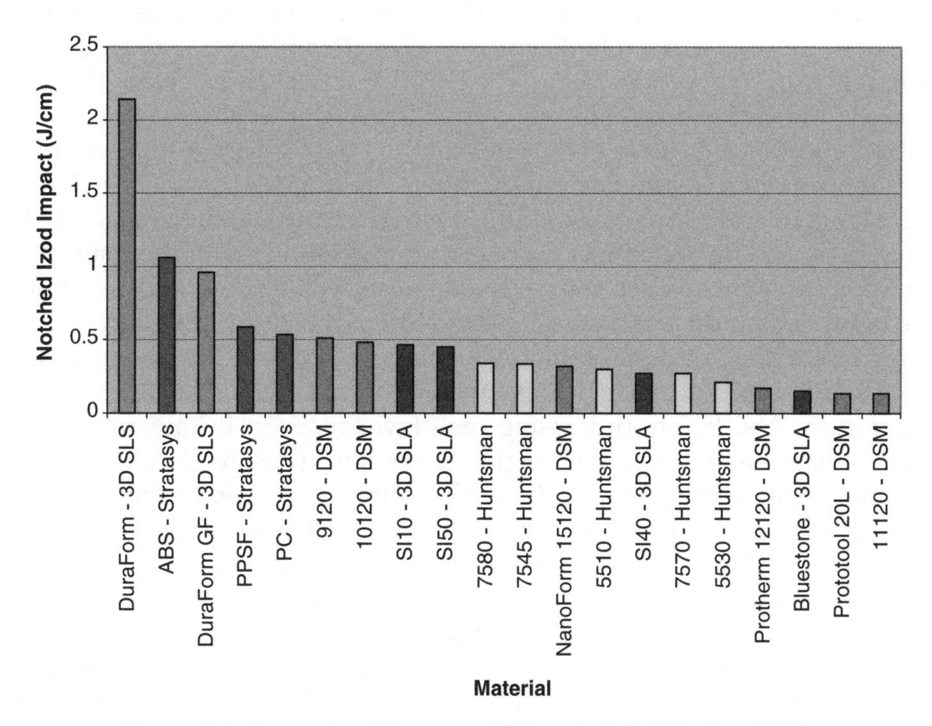


**Figure 8.4** Impact resistances of RP materials (Mueller, 2004)

**Table 8.1** SOMOS 8110 and 11120 mechanical properties 15 minutes per side UV post cure

| DSM SOMOS material | Axis of ASTM bar | Yield strength (ksi) | Tensile strength (ultimate) (ksi) | Elongation at break (%) |
|---|---|---|---|---|
| SOMOS 8110 | X | 1.3 | 3.8 | 17.6 |
| | XZ (on edge) | 1.4 | 4.0 | 17.3 |
| | Z | 1.4 | 3.8 | 15.0 |
| SOMOS 11120 | X | 5.3 | 7.5 | 5.0 |
| | XZ (on edge) | 3.8 | 7.6 | 4.0 |
| | Z | 4.7 | 7.7 | 5.2 |

**Table 8.2** SOMOS 8110 and 11120 mechanical properties from the vendor

| DSM SOMOS material | Tensile strength (ksi) | Elongation at break (%) |
|---|---|---|
| SOMOS 8110 | 2.6 | 27 |
| SOMOS 11120 | 6.8–7.8 | 11–20 |

that the vendor data states their values as approximate and may vary upon build conditions.

The tensile bars for the SOMOS 8110 as built were very consistent in all directions but had a much higher tensile strength and lower elongation than expected from the vendor data. The SOMOS 11120 had consistent tensile strength and elongations, but more variation in the yield strength.

The importance of these data is not that the tensile bars tested differently from the vendor's stated data but that there are important variables that can affect the part and its mechanical properties that must be understood if SL is going to be an effective direct manufacturing method. With the maturity of the SL equipment the critical parameters are well known and monitored. The resin activation parameters, provided by the resin manufacturer, are entered into the machine and the machine calculates the required laser powers and draw speeds to achieve the optimum results. The laser power is tested at the bed before processing each layer to assist in this. In addition, build 'styles' may be provided to give the best part results for a given resin.

The data sheets provide very little information, if any, on the post cure that was applied to the parts to achieve mechanical properties. This can be through the UV post cure in an oven or sunlight, or a thermal cure. Varying the UV post cure time can affect mechanical properties. Thermal curing is often used to increase the HDT, but it also affects the mechanical properties.

Another issue with SL epoxy acrylate thermoset resins is the stability of the mechanical properties over time. It is a well-known phenomenon that due to the physical chemistry of the material its properties continue to

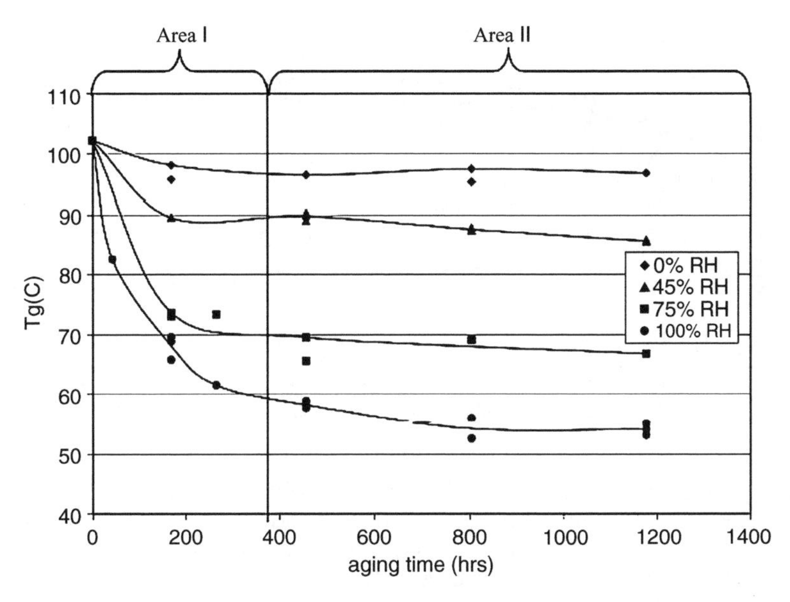


**Figure 8.5** Glass transition temperature of SL7510 SLA resin. (From X. Ottemer and J.S. Colton, Effects of aging on epoxy-based rapid tooling materials, *Rapid Prototyping Journal*, 2002, **8**(4), 215–23 © MCB UP Limited; http://www.emeraldinsight.com/rpj.htm. Republished with permission, Emerald Group Publishing Limited)

change over time. Heat and humidity can accelerate loss of mechanical properties which can cause warping and failure of parts if loaded.

Work by Ottemer and Colton (2002) on aging of epoxy-based SL parts in various relative humidities showed that water absorption affected both glass transition temperature and mechanical properties. Water breaks crosslinks between molecules causing plasticization of the resin. However, after an initial 400 hour induction period for equilibrium of the moisture, the glass transition temperature ($T_g$) and mechanical properties stabilized. Figure 8.5 illustrates the effect of relative humidity (RH) on the SL7510 SLA resin's $T_g$. Young's modulus ($E_y$) exhibits a similar phenomenon, decreasing as the relative RH increases, as shown in Figure 8.6. In the dry environment, $E_y$ has a slight increase through the induction timeframe.

Some of the newer resins are being tailored to reduce water absorption and the effect of residual UV exposure, which hopefully can assist in improving this issue. However, little vendor-supplied data or publicly presented information exists on this subject.

### *8.2.1 Viability for Series Rapid Manufacturing*

The SL-based equipment is still considered to be the most accurate of the additive manufacturing equipments, with many platform sizes and build

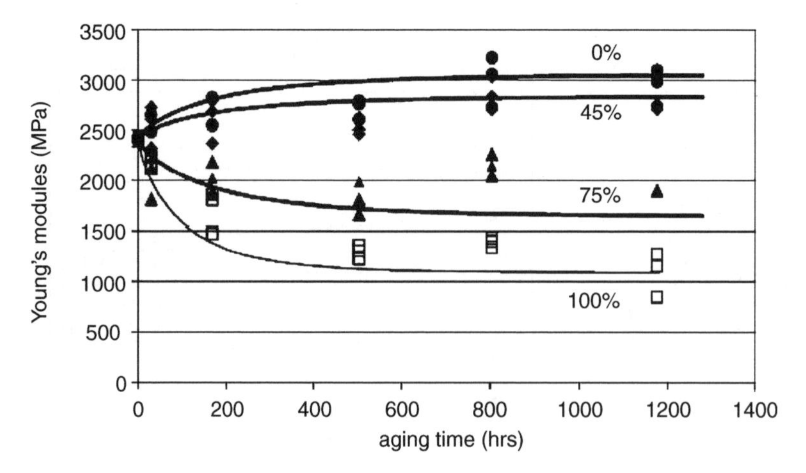


**Figure 8.6** Young's modulus for SL7510 SLA resin. (From X. Ottemer and J.S. Colton, Effects of aging on epoxy-based rapid tooling materials, *Rapid Prototyping Journal*, 2002, **8**(4), 215–23 © MCB UP limited; http://www.emeraldinsight.com/rpj.htm. Republished with permission, Emerald Group Publishing Limited)

speeds available. The large and varied material systems available offer an end user many choices in mechanical properties. The mechanical properties appear to be less orientation specific (anisotropic) and more consistent. However, the issue of long-term stability, especially for structurally loaded geometries, currently limits its application for end-use serial parts with the current range of materials. Post curing (UV and/or thermal) are variables that must be controlled to achieve a repeatable process for manufacturing. New resin developments, possible with the urethane-based material systems, could offer new opportunities if they can be proven to increase long-term stability.

## 8.3 Selective Laser Sintering

The advantage of selective laser sintering (SLS) has always been its ability to utilize functional engineering thermoplastics in the equipment. The powdered plastic is preheated to just below the melting point of the material. The laser then only adds an incremental amount of power to the powder to melt it into the desired cross-section.

The SLS process is best suited to semi-crystalline to crystalline polymers as they exhibit a sharp melting point. Figure 8.7 is a differential scanning calorimetry plot of nylon-12, exhibiting the relatively sharp melting point. Amorphous polymers such as polycarbonate have a softening range that is somewhat broad. These materials are not well suited to SLS for Rapid Manufacture as it is difficult to get a complete melt of the material.

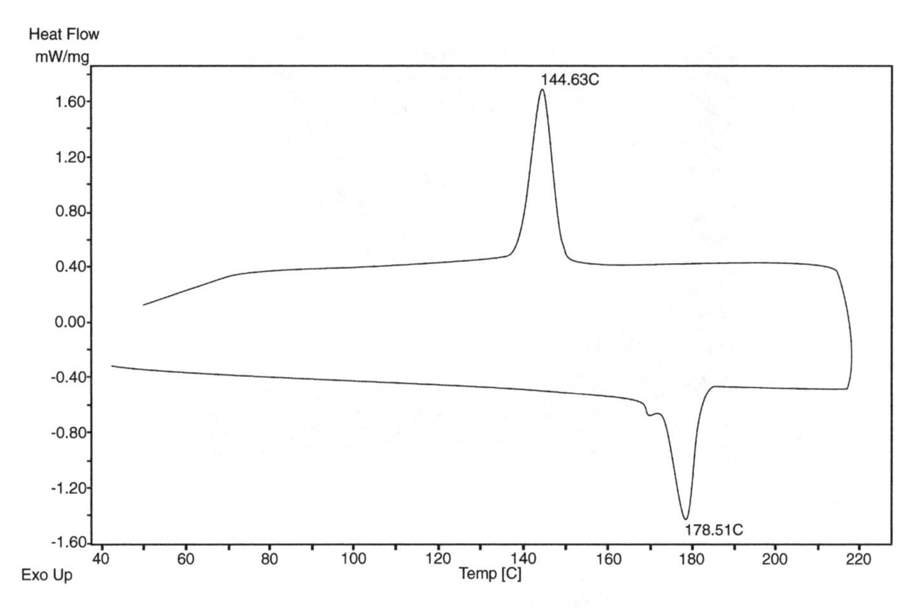


**Figure 8.7** Differential scanning calorimetry plot

Polycarbonate was initially offered for the SLS process, but as a casting pattern, not a functional material system. The parts created with amorphous materials with the SLS process tend to be very weak and brittle as the powder does not fully melt. The interior of the part appears more like Styrofoam as it is mostly powder beads that are joined together, but not fully melted. The material will 'burn' if too much laser power is applied.

In injection molding, material is subjected to a full melt as it goes through the barrel and is injected into the mold under high pressure. The gate and runner system is often re-ground into pellets and put back into the material stream as 're-grind'. This reduces costs by reducing the amount of virgin plastic that must be used. However, the multiple heating cycles degrade the polymer and impact the mechanical properties of the production part. For most parts, the re-grind percentage is limited to a specific amount so as not to affect the final part.

The SLS process also subjects the powders to a thermal history near the melt point as it is heated both in the feed cartridges and in the build area. Thus, material quality is an issue as well. It is well known within the SLS user community that the polyamide powders are not 100 % recyclable. The continuous use of powder without refreshing with virgin powder can have a negative effect on both part quality and mechanical properties. With the polyamide (nylon) powders, the surface finish can become very rough, a condition known as 'orange peel', as seen in Figure 8.8. For DuraForm$^{TM}$, a blend of one-third virgin material to two-thirds overflow and part cake is normal. The glass-filled (GF) material is generally blended one-half virgin to

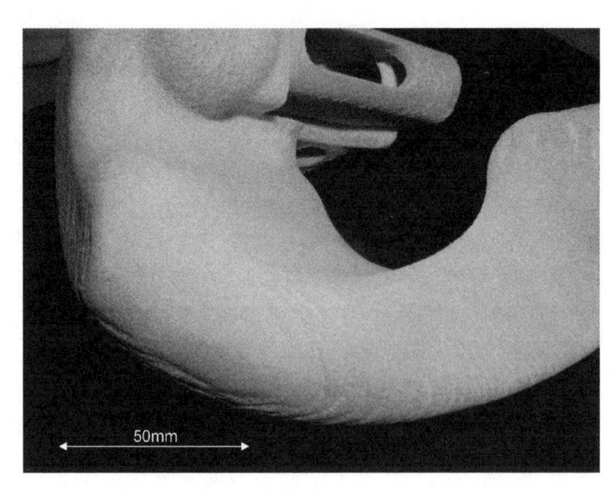


**Figure 8.8** Surface roughness or 'orange peel' of a DuraForm GF$^{TM}$ part

one-half used as its properties, especially surface finish, tend to degrade more quickly.

A study of the DuraForm$^{TM}$ and glass-filled DuraForm$^{TM}$ material systems performed at the University of Louisville, Davis, Gornet and Richardson (2003), was run to determine the effect of material quality on mechanical properties. The initial run used virgin powder only and ran the feed cartridges empty. Therefore, all the powder was exposed to a thermal history. Each successive run used all the powder from the part cake as well as the overflow cartridges. Laser power, part bed temperature and feed cartridge temperature were held constant. Due to the loss of material from the sintered parts, each successive build would have less powder available so it would be a shorter build. This was continued until parts could no longer be built or part quality degraded to an unacceptable level. ASTM D638 tensile bars, scale measurement parts and various part geometries for surface quality evaluation were included in the build.

The ASTM D638 tensile bars from each build were saved and tested. In addition, powder samples were saved from each build. The tensile and elongation results are shown in Figure 8.9. The melt point and melt flow rate (MFR) are shown in Figure 8.10. These results are for bars oriented in the $X$ and $Y$ directions. The tensile strengths remained consistent until the sixth build, where properties started to fall quickly. Interestingly, the elongation at break increased with successive builds. The decrease in MFR shown in Figure 8.10 is indicative of an increase in molecular chain length, which causes the observed increase in melt temperature for successive builds. It was also noted that the scale and offsets also changed with each build.

The MFR was measured using an extrusion plastometer according to ASTM D1238. This index is a measure of the flow characteristic of the molten

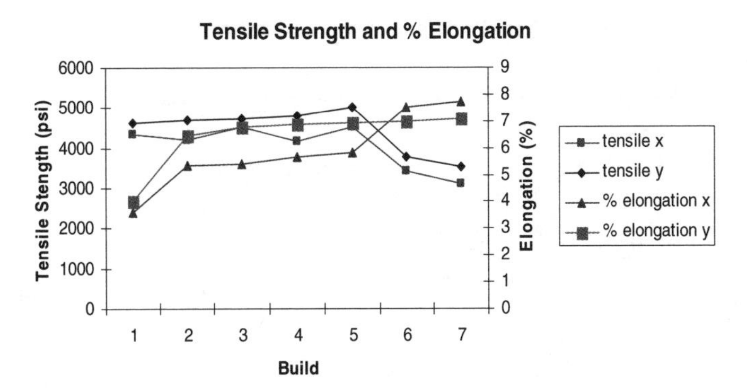


**Figure 8.9** Tensile strength and elongation for DuraForm™

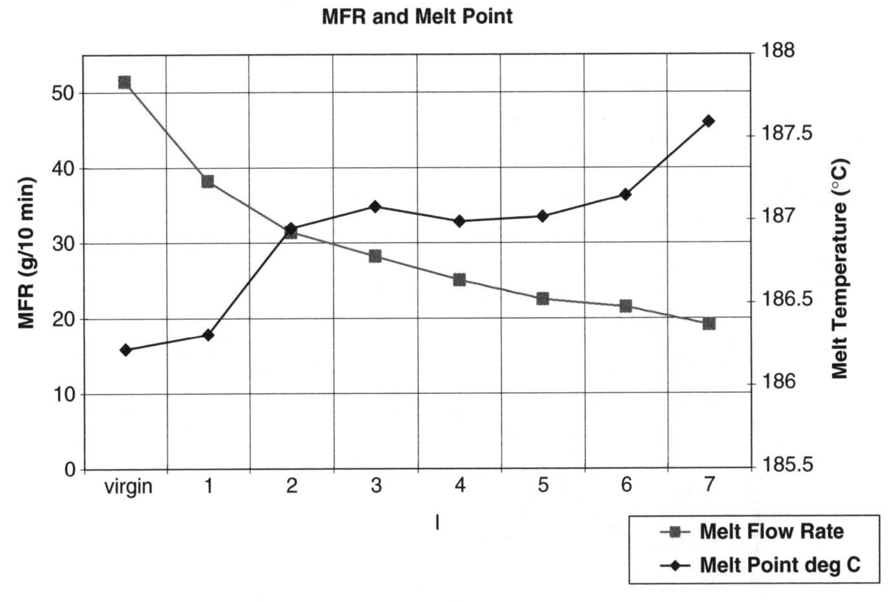


**Figure 8.10** Melt point and melt flow rate for DuraForm™

polymer and is sensitive to differences in the basic polymer structure due to changes in molecular weight. Changes in the melt index should correlate with changes in the build characteristics of DuraForm™ during the laser sintering process.

The melt index test is a standard test in injection molding to evaluate material degradation that could be caused by a high barrel residence time or a large percentage of re-grind material in the blend. In general the melt index increases as the material degrades, indicating a decrease in the molecular weight. Contrary to injection moulding, the results for DuraForm™ show the MFR decreasing with successive use.

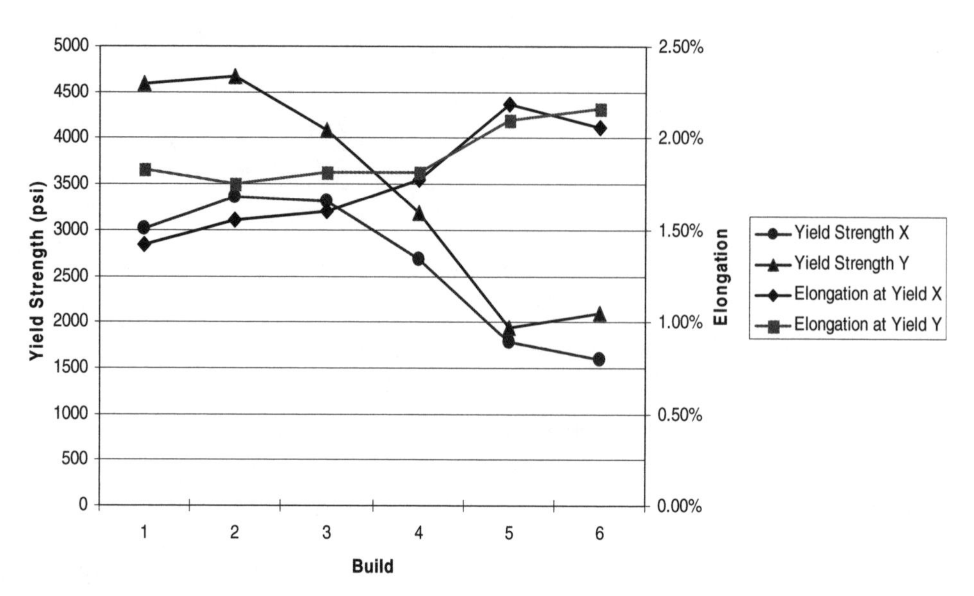


**Figure 8.11** Tensile strength and elongation for glass-filled Duraform™ runs

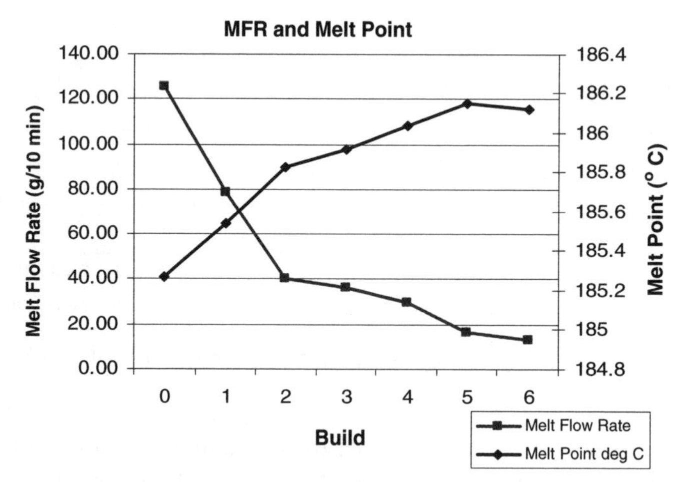


**Figure 8.12** Melt point and melt flow rate of glass-filled DuraForm™ runs

Figures 8.11 and 8.12 show the same results as those shown in Figures 8.9 and 8.10 but for glass-filled DuraForm™. The mechanical properties degrade at a much faster rate than the unfilled material. As material ages it is also necessary to raise the temperature of the build area to compensate for the increased melt point. The glass-filled DuraForm™ material shows a more rapid decrease in the melt flow rate. The trend of the melt point curve indicates that the melt point is increasing with each successive heat exposurein the SLS machine. The combination of these two results points to a molecular weight increase of the polymer when subjected to multiple heating cycles.

Direct manufacturing requires that mechanical properties and part accuracy are consistent in order to create a part with an identifiable quality. For the DuraForm™ polyamide materials, utilizing a material of a consistent quality should make the build temperature, scale and offset, and mechanical properties constant, resulting in a more controllable process.

The University of Louisville Rapid Prototyping Center developed a novel material blending method based on the melt flow rate to minimize process variations due to material degradation and guarantee physical properties and dimensional and surface finish characteristics. Using the MFR as the metric, users can blend powder to a specific target melt index using blending curves to keep the material quality static. This eliminates the effect that hot and cold builds have on traditional blending. It has been installed at several industrial sites where it has stabilized machine operating parameters and increased consistency in part quality and strength. It can also be used for incoming inspection of material and process troubleshooting.

Part orientation also plays a critical role in mechanical properties via SLS. Parts constructed via SLS exhibit better strength and elongation when built in the *X* and *Y* axis, the plane of laser motion. These are the results usually reported by the vendor. Parts built oriented along the *Z* axis have a significantly lower elongation. In high-pressure injection molding, the molecules align with the direction of flow. This is especially true for glass fiber filled materials. In the SLS process there is no applied pressure to 'pack' the material and the materials are glass bead filled instead of fiber filled. Table 8.3 shows the average results recorded from SLS samples built in different orientations. These results show the effects of build orientation on SLS parts indicating a high degree of anisotropy.

Table 8.4 shows the material properties for SLS Duraform™ and glass-filled DuraForm™ as quoted by 3D Systems. The discrepancies between the values presented in Tables 8.3 and 8.4 indicate that process control as

**Table 8.3** Selective laser sintering DuraForm™ and DuraForm GF™ as built properties

| | Axis of ASTM bar | Yield strength (ksi) | Tensile modulus (ksi) | Ultimate strength (ksi) | Elongation (%) |
|---|---|---|---|---|---|
| DuraForm™ | *X, Y* | 3.9 | 264 | 6.6 | 8.0 |
| DuraForm™ | *ZX,ZY* | 4.5 | 281 | 6.3 | 4.0 |
| DuraForm GF™ | *X,Y* | 3.9 | 439 | 5.9 | 3.0 |
| DuraForm GF™ | *ZX,ZY* | 4.8 | 548 | 6.4 | 2.2 |

**Table 8.4** DuraForm™ and DuraForm GF™ properties (3D Systems)

| Material | Tensile modulus (ksi) | Tensile strength (ksi) | Elongation (%) |
|---|---|---|---|
| DuraForm™ | 232 | 6.4 | 9 |
| DuraForm GF™ | 857 | 5.5 | 2 |

well as orientation play an important role when trying to achieve consistent mechanical properties to satisfy design requirements for series RM.

### *8.3.1 Viability for Series Rapid Manufacturing using SLS*

The selective laser sintering process currently has the deepest penetration into the direct serial manufacturing realm, with highly successful projects with Boeing and Siemens/Phonak due to the polyamide engineering thermoplastics material systems. Monitoring of material quality and machine consistency aides in the process control aspects of this technology. 3D Systems has released a new system and upgrade called the HiQ system which monitors and calibrates the infrared sensors in the machine during a build. This new development increases thermal control of the machine increasing its functionality for manufacturing by reducing machine to machine variations.

Increasing the material system capabilities by offering materials with better mechanical properties, higher application temperatures and increased impact strength could bring the utilization of the SLS process into direct serial manufacturing into a new realm.

## 8.4 Fused Deposition Modeling

Fused deposition modeling (FDM) extrudes a thermoplastic filament through a heated tip to create parts (see Chapter 5). This process offers the unique ability of melting both crystalline and amorphous polymers in the system. The material approaches a full melt during the extrusion. Another tip extrudes support material that can be removed physically or dissolved in a waterbath. The currently available materials are ABS, polycarbonate (PC) and polyphenylsulfone (PPSF). These engineering thermoplastics offer a broad range of functionality. Stratasys currently does not provide mechanical properties of their materials in an as-built state.

In general, an outline of the perimeter of the part is extruded from the head and then the interior is raster filled by the extruder head. Different tip diameters are available depending upon the resolution and layer thickness required. The orientation of the raster scans can be controlled as well. The default is filling with a $45/-45^\circ$ raster scan. Other options include in the $X$

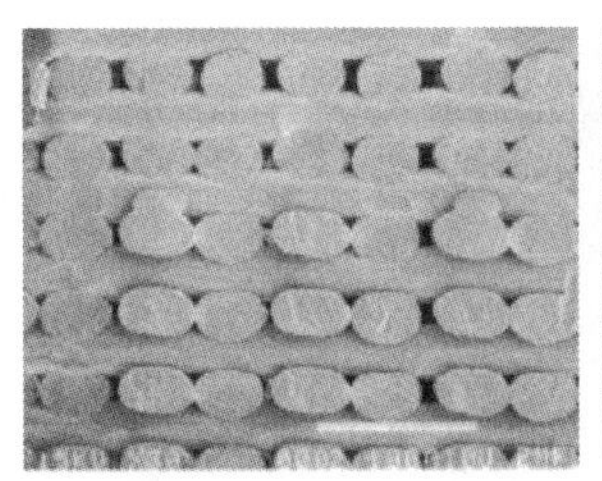

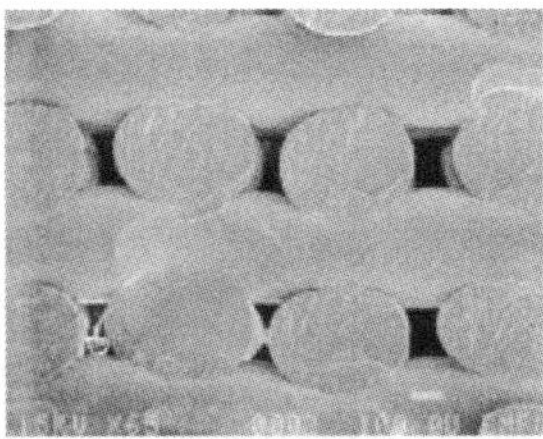

**Figure 8.13** Fused deposition modeling cross-section with a 0/90° raster. (From A. Bellini and S. Guceri, Mechanical characterization of parts fabricated using fused deposition modeling, *Rapid Prototyping Journal*, 2003, **9**(4), 252–64 © MCB UP Limited; http://www.emeraldingsight.com/rpj.htm. Republished with permission, Emerald Group Publishing Limited)

direction only, $Y$ direction only and a 0/90° cross-scan. The density of the interior cross-hatching can be controlled to decrease the fill amount, thereby reducing build time and part density via a positive air gap setting. The cross-hatching can also be tightened to create a more dense hatching by slightly overlapping the extruded roads to create a more dense part with better mechanical properties. The temperature of the extruder tip can also be varied if required.

Several previous studies of the mechanical properties of the FDM process have been offered by Ahn *et al.* (2002) and Bellini and Guceri (2003), with very interesting results. Figure 8.13 is a cross-sectional view of an ABS part built with 0/90° raster patterning. From this view one can see each individual deposited road width. Each deposited road sticks to the previously deposited layer. However, since the outer surface of the extruded material cools fairly rapidly, there is no mixing of material from layer to layer or road width to road width.

This is similar to injection molding when an injected flow of material splits and has to reform or meet back-up. If the material is warm enough and moving quickly, it can reform into a single front. If not, the split flow front forms what is known as a cold weld or knit line. The material does not mix across the fronts and a singularity exists. This is very true of fiber reinforced material. This knit line is solid, but acts as a stress riser or localized area of lower mechanical properties.

Ahn *et al.* (2002) injection molded the FDM material into ASTM tensile bars to provide a baseline tensile strength of 26 MPa. FDM samples were prepared with the cross-hatching in all four primary orientations: axial along the length of the bar, 45/−45°, 0/90° and transverse across the bar. Two populations of bars, one with a zero air gap or no road width overlap, and a −0.003 inch air gap, were built and tested.

Figures 8.14 and 8.15 illustrate the resultant tensile strengths. The strengths range from 10 to 73 % of the injection molded samples. The default

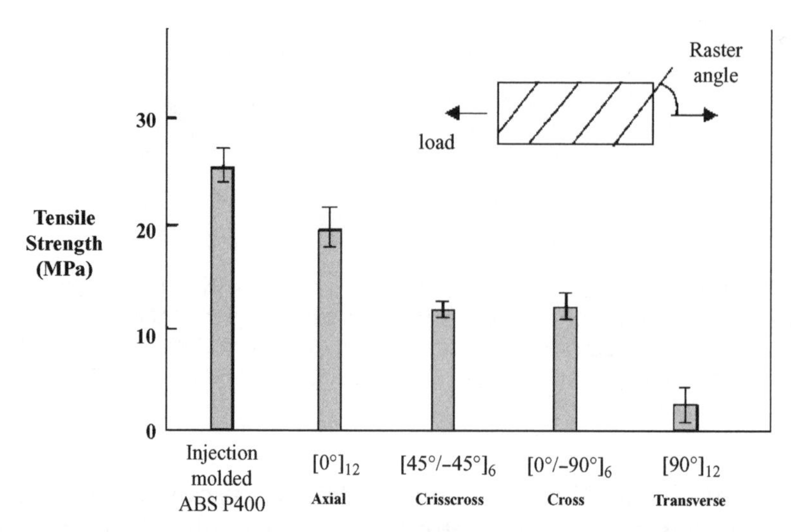


**Figure 8.14** Tensile strength of zero air gap FDM ABS P400 samples. (From S. Ahn, M. Montero, D. Odell, S. Roundy and P.K. Wright, anisotropic properties of fused deposition modeling ABS, *Rapid Prototyping Journal*, 2002, **8**(4), 248–57 © MCB UP Limited; http://www.emeraldinsight.com/rpj.htm. Republished with permission, Emerald Group Publishing Limited)

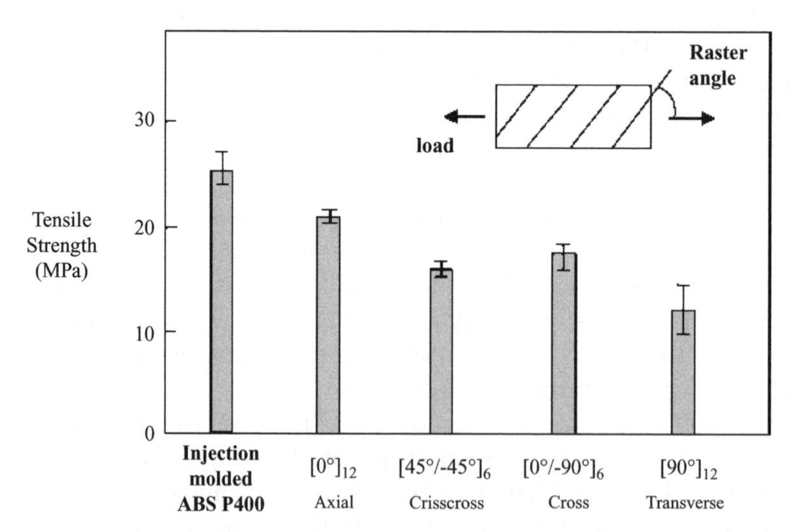


**Figure 8.15** Tensile strength of −0.003 inch FDM ABS P400 samples. (From S. Ahn, M. Montero, D. Odell, S. Roundy and P.K. Wright, anisotropic properties of fused deposition modeling ABS, *Rapid Prototyping Journal*, 2002, **8**(4), 248–57 © MCB UP Limited; http://www.emeraldinsight.com/rpj.htm. Republished with permission, Emerald Group Publishing Limited)

**Table 8.5** Tensile test results. (From A. Bellini and S. Guceri, Mechanical characterization of parts fabricated using fused deposition modeling, *Rapid Prototyping Journal*, 2003, **9**(4), 252—64 © MCB UP Limited; http://www.emeraldingsight.com/rpj.htm. Republished with permission, Emerald Group Publishing Limited)

| Orientation | Tensile strength (MPa) |
|---|---|
| *X* | 11.7 |
| *Y* | 16.0 |
| *Z* | 7.6 |

scan pattern, 45/−45°, is about 50 % of the injection molded samples. As one would expect, the sample with the road widths running the length of the tensile bar achieves the best result while the one running across the bar achieves the lowest. The negative air gap samples improved the part density and mechanical properties in all cases. The default scan pattern of 45/−45° improved the tensile strength to 65 % of the molded part properties. Clearly, the FDM construction method results in anisotropic properties.

Bellini and Guceri expanded the knowledge base by including parts oriented in principle directions other than along the *X* axis of the machine envelope. Initially, mechanical tests were run on the FDM filament and on single roads extruded on the FDM machine. The FDM filament had a tensile strength of 34.3 MPa and an elongation at break of 53 %. The individual road had a similar tensile strength but the elongation dropped to 16.9 %. To minimize the orientation of the cross-hatching, a combination of 0/90° and 45/−45° was run. Table 8.5 shows the tensile strength of ABS parts made by FDM at different orientations indicating a high degree of anisotropy.

Again, anisotropic mechanical properties were exhibited for the FDM ABS material. The weakest orientation is the *Z* direction, owing to the individual layers running through the part with no intermixing of material between each layer.

There is not a sufficient body of knowledge of mechanical properties for the Stratasys materials such as yield strength, tensile strength, flexural modulus and elongation to completely characterize the FDM process for direct manufacturing at this time.

### *8.4.1 Viability for Series Rapid Manufacturing*

The fused deposition modeling process is exciting in the fact that it has the capability of using a wide range of crystalline and amorphous thermoplastics. To best take advantage of the material properties the parts should be built with a negative air gap to maximize part density and mechanical properties. This increases build time. However, due to the extrusion process and lack of material intermixing, the resultant parts will not achieve the full mechanical properties of the base thermoplastic. The raster scan patterns assist in creating an anisotropic property condition. The interlayer bonding issue will cause the *Z* axis mechanical properties to be the deciding design factors.

It is important to evaluate the different build styles among the various materials to determine the behavior of each material. Additional mechanical property work could assist in developing better design guidelines for series Rapid Manufacturing.

## 8.5 Metal-Based Processes

The systems that are able to melt, deposit or bond molten metals without a secondary infiltration process have the best opportunity for direct manufacturing without a sacrifice in mechanical properties. Most of these systems utilize powdered metals and are selectively melted in a powder bed or the powder is fed into a laser beam, where it is melted and deposited. Common materials seen in these processes are tool steels and titanium.

### *8.5.1 Fused Metal Deposition Systems*

The two commercial systems available today are the Optomec laser engineered net shaping (LENS) and the POM direct metal deposition (DMD). The DMD and LENS systems feed the powder into the laser through a nozzle on to a substrate with up to five axes of control. They can be used for either direct creation of a part or tool or for repairing or reconfiguring of a part or a tool. In both cases one achieves a metallurgical bond as opposed to the mechanical bond of a weld. The laser acts as a mixing device to melt some of the previous layer as it deposits. They can deposit pure metals, e.g. H13 tool steel, which allows for repair of molds that require a high polish without resulting in a visual defect in the molded part. Figure 8.16 shows a polished sample of an H13 deposit via DMD. Notice that the only difference is the finer grain structure of the DMD deposit compared to the wrought H13 substrate.

These machines act as 'microfoundries' due to the controlled quenching achieved by the very localized heating and deposition. The hardness of some materials can be controlled during the deposition, allowing one to match an existing part without the requirement for heat treating. These systems have multiple powder feed cartridges that give the unique opportunity for creating multiple material or gradient structures where the composition can be changed in three dimensions (see Chapter 7). In addition, ceramics or other non-metallic materials can be added through one of the feeders to offer localized property enhancement for wear or cutting surface properties.

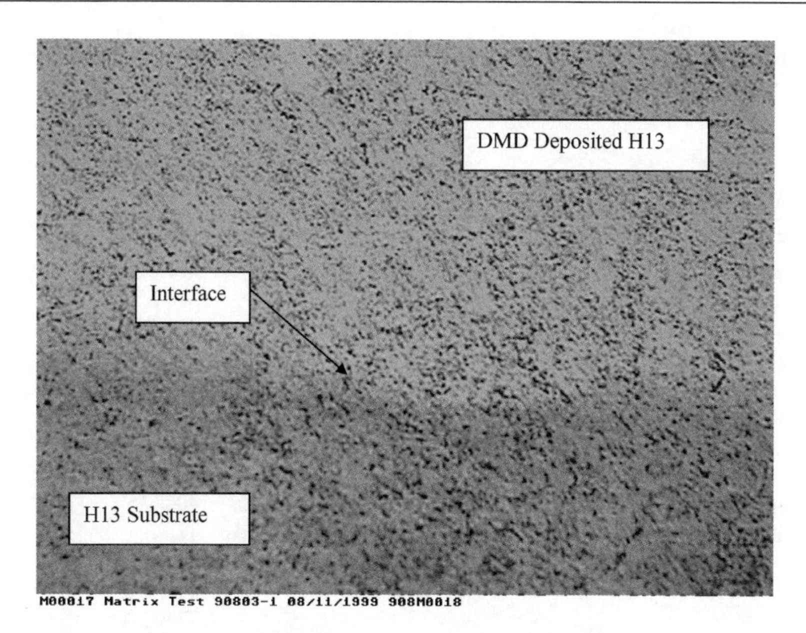


**Figure 8.16** Micrograph of direct metal deposition of H13 on to an H13 substrate. (Reproduced by permission of POM)

By adding different materials to each other via these methods can allow one to take advantage of dissimiliar materials in an environment where one or the other would not normally be used. High thermal conductivity tooling is a recent application that utilizes a machined copper preform and applies a hardened tool steel surface. Copper is normally too soft to be used as a tool surface material, but by having an H13 hardened surface one can get the best of both materials in a single part. Figure 8.17 shows a high thermal

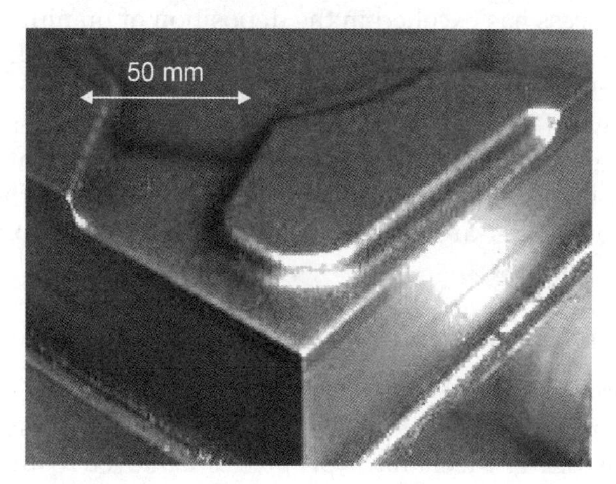


**Figure 8.17** High thermal conductivity tool by direct metal deposition (University of Louisville)

**Table 8.6** Materials used by current fused metal deposition systems

| | Alloy | |
|---|---|---|
| Material | Direct metal deposition | Laser engineered net shaping |
| Steels | H13 | H13 |
| | H19 | 304 SS |
| | P20 | 316 SS |
| | P21 | 420 SS |
| | 420 SS | S7 |
| | S7 | 17-4 PH |
| | D2 | |
| | 17-4 PH | |
| | 15-5 PH | |
| Aluminum | | 4047 |
| Titanium | Ti | Ti |
| | Ti 6-4 | Ti 6-4 |
| | | Ti 6-2-4-2 |
| | | Ti 6-2-4-6 |
| Nickel based | Inconels | Inconels |
| | Nickel super alloys | Hastelloy X |
| Cobalts | Stellite | Stellite |
| Copper | Cu | Cu–Ni |

conductivity tool portion that is able to reduce injection molding or die casting cooling times by 20–40 %. The multiple material feeders can make it possible to vary electrical and thermal conductivity, wear properties, hardness and mechanical strength throughout a part.

The LENS process has excelled in the deposition of titanium and its alloys. Operating in a vacuum environment it can deposit titanium and achieve equivalent mechanical properties to that of a cast or wrought alloy.

Table 8.6 shows a chart of the current types of materials deposited by each system. There are numerous other alloys under development for LENS and DMD by equipment manufacturers and probably more by their customers. The majority of metals are readily available in powder form from other manufacturing processes.

### *8.5.2 Viability for Series Rapid Manufacturing*

Both the Optomec laser engineered net shaping (LENS) and the POM direct metal deposition (DMD) technologies offer very good prospects for series Rapid Manufacturing. Mechanical properties achieved with these systems meet or exceed wrought alloy standards. The LENS process has shown particular strength in titanium and its alloys, which is applicable to the medical field, defense and aerospace. The DMD process has concentrated on the tool steels. Both technologies offer the promise of gradient structures to enable totally new mechanical properties in a smart structure.

The current negative to these systems is a relatively low deposition rate, making larger parts less cost effective. Both systems have a five-axis deposition head available but there must still be a line of site deposition and space for the head to maneuver. At this time there are no support materials so overhanging geometry is an issue. The fact that these processes provide numerous material options including compositional gradients will mean that characterization and achieving repeatability of properties will be an onerous task.

### *8.5.3 Powder Bed Systems*

These systems are basically selective laser sintering but a full melt of a metal powder is achieved in the bed. The electron beam melting (EBM) system from Arcam, Sweden, uses an electron beam to melt the metallic powders. The selective laser melting (SLM) of MCP-HEK, Germany, uses an infrared laser in its processing. The advantages to the powder bed systems are that support structures are often not required and there are many powder options. The disadvantage is that they currently cannot build from more than one powder at a time. High density is achieved and similar mechanical properties to the wrought alloys have been seen. Table 8.7 shows the materials that are available for the Arcam and MCP-HEK processes.

Currently a number of other metal-based powder bed processes are available, including the well-established EOS direct metal laser sintering and the 3D Systems indirect metal laser sintering along with newer processes such as Trumpf's direct laser forming.

### *8.5.4 Ultrasonic Consolidation*

Ultrasonic consolidation (UC) is a technology from Solidica, Inc., USA. It uses aluminum tape that is bonded to the previous layer via ultrasonic

**Table 8.7** Materials for the Arcam and MCP-HEK processes

| Arcam electron beam melting | MCP-HEK selective laser melting |
|---|---|
| Titanium 6-4 | Zinc |
| H13 tool steel | Bronze |
| | Stainless steel |
| | Tool steel |
| | Cobalt chromium |
| | Ti |
| | Silicone carbide |
| | Aluminum oxide |

welding. The shape is then machined so that it is a combination additive/subtractive process. Currently aluminum is the only material available. Its mechanical properties after construction are similar to wrought alloys.

### 8.5.5 Viability for Direct Serial Manufacturing

The powder bed systems from Arcam, Trumpf and MCP-Hek as well as the Solidica ultrasonic consolidation offer single material depositions. They are not available for gradient or even multiple material capabilities at this time. Many of the metal-based powder bed sintering systems require supports; for example, in the EOS direct metal laser sintering process supports are required to anchor overhangs and in Arcam's EBM process supports are required to provide a good electrically conductive passage through the parts to the build platform. This is likely to reduce the complexity of parts that can be produced compared with, for example, polymer laser sintering, but should still allow the production of parts that cannot be done by conventional processes such as machining.

Current mechanical property data for the metal-based processes are good but are very limited at this time. As with the more mature sintering of polymers, repeatability is likely to be crucial for facilitating more widespread adoption of these processes for series Rapid Manufacturing.

## References

1. Ahn, S., Montero, M., Odell, D., Roundy, S. and Wright, P.K. (2002) Anisotropic properties of fused deposition modeling ABS, *Rapid Prototyping Journal*, **8**(4), 248–57.
2. ASTM D638-03 (2003) *Standard Test Method for Tensile Properties of Plastics*, American Society for Testing and Materials, Philadelphia, Pennsylvania.
3. Bellini, A. and Guceri, S. (2003) Mechanical characterization of parts fabricated using fused deposition modeling, *Rapid Prototyping Journal*, **9**(4), 252–64.
4. Davis, K.R., Gornet, T.J. and Richardson, K.M. (2003) Material testing method for process control of direct manufacturing in the SLS process, in Proceeding of the 1st International Conference on *Advanced Research in Virtual and Rapid Prototyping*, Leiria, Portugal, 1–4 October 2003.
5. Mueller, T. (2004) Truly functional testing; selecting rapid prototyping materials so that prototypes predict the performance of injection molded plastic parts, in Rapid Prototyping and Manufacturing Conference, Society of Manufacturing Engineers, Dearbon, Michigan, 10–13 May 2004.
6. Ottemer, X. and Colton, J.S. (2002) Effects of aging on epoxy-based rapid tooling materials, *Rapid Prototyping Journal*, **8**(4), 215–23.

# 9

# Production Economics of Rapid Manufacture

Neil Hopkinson
*Loughborough University*

## 9.1 Introduction

Understanding the economic implications of Rapid Manufacture is critical for any organisation considering adopting the technology. The economics of Rapid Manufacture falls into two distinct categories, which may be described as direct production economics and manufacturing system economics. This chapter considers the direct production costs associated with creating physical products from computer aided design (CAD) files; this is a relatively simple subject and gives quantifiable results that indicate the suitability or unsuitability of adopting Rapid Manufacture by comparing costs with alternative processes. The more complex field of manufacturing system economics is covered in the next chapter on Management and Implementation of Rapid Manufacturing. Any organisation that is considering the adoption of Rapid Manufacture will need to consider both aspects of the economics involved, but the production side may best be considered first due to its simplicity and the ability to make judgements based on hard data. It should be stressed that this chapter considers series manufacture (the same parts again and again) on existing commercial (Rapid Prototyping) equipment. However, many of today's and the future's most successful applications of Rapid Manufacture involve the manufacture of customised products in low or unit production volumes. Also, technologies that are

*Rapid Manufacturing: An Industrial Revolution for the Digital Age*
Editors N. Hopkinson, R.J.M. Hague and P.M. Dickens 

more suited to high-volume production (see Chapter 5 on Emerging Rapid Manufacturing Processes) will have a significant impact on the production economics of Rapid Manufacture.

Probably the most important message to be conveyed in this chapter, especially for readers with a background in Rapid Prototyping (RP) and Rapid Tooling (RT), is that the costs associated with Rapid Manufacture are not the same as those associated with RP. This is often a difficult concept to accept – after all both involve making parts from CAD files on the same machines. Some of the results presented below received a very mixed response when first presented; in particular, many RP practitioners could not envisage a viable business plan based on making parts for very low cost (a few euros) on RP machines. However, as we will go on to see, production of low-cost parts on RP machines is not only possible but also absolutely necessary for some organisations to become and remain successful.

The production economics of Rapid Manufacture considers the cost of converting a CAD file (.STL format) into a series of finished products. In order to do this, many of the rules applied to conventional manufacturing processes are applied to Rapid Prototyping/Manufacturing (RP/M) processes. For example, it is assumed that a machine is used to produce the same part again and again – something that would not apply for RP. The breakdown of costs to produce parts by Rapid Manufacture fall into machine costs, material costs and labour costs. The issue of overheads is not considered here as it forms a central part of system economics that is discussed later.

## 9.2 Machine Costs

Machine costs for Rapid Manufacture are largely dictated by machine depreciation. Manufacturing capital equipment is normally depreciated using a straight line over 8–10 years (7–10 years in the USA) [1] and so for Rapid Manufacture these rules apply, representing a significant departure from costing for RP, where depreciation is frequently applied over a shorter timescale, typically 5 years. Using an extended period for depreciation is valid so long as the equipment remains functional for that time period. This brings us on to the second factor of machine cost – maintenance. Using a selective laser sintering (SLS) machine as an example for Rapid Manufacture, the equipment will generally remain functional over an extended timescale so long as parts (notably lasers) are replaced as and when required. In order to ensure this, Rapid Manufacturing (RM) costs need to factor in a full maintenance package including the replacement of lasers as and when required for the useful life of the machine. Machine utilisation is another area of discrepancy between costs for RM and RP. In an RP scenario, a machine is likely to be in use for typically 60 % of the time as the setting up and removal of unique parts and builds is time consuming and builds often finish at times when there are no personnel present to remove them and begin a next build. However, in an RM environment, where identical builds with known build times are repeated, utilisation is much improved. For example, the RM company ODM in California regularly achieves machine utilisation figures in the region of 85 %. Increasing the useful life and percentage utilisation of a machine combines to significantly reduce the cost of parts attributed to the machine. The requirement for a full maintenance package may incur an increase in price but this is typically insignificant when compared to the cost reduction of using RP equipment for Rapid Manufacture.

## 9.3 Material Costs

The price per kilogram of materials for Rapid Manufacture are far higher than those for conventional manufacturing processes. Table 9.1 shows the approximate costs of materials for various RP processes compared with those for injection moulding and machining.

These figures suggest that RM will struggle to compete with conventional processes where high material volumes through high product weight and/or high production volumes are involved. However, with the increased use of Rapid Manufacture (and to some extent RP/T) larger volumes of material are being used and ultimately this should pave the way for lower material costs due to economies of scale. Current costs of materials are not simply high due to economies of scale; research and development costs and production costs will invariably result in higher costs for Rapid Manufacture than for conventional processes for some time to come.

One of the perceived benefits of Rapid Manufacture is the potential to create products with zero waste. However, in reality this is not the case. Most processes require support structures (stereolithography, fused deposition

**Table 9.1** Approximate material costs for different processes (1–3)

| | Process/material | Cost per kg ($) |
|---|---|---|
| Rapid prototyping/manufacturing | Stereolithography/epoxy-based resin | 175 |
| | Selective laser sintering/nylon powder | 75 |
| | Fused deposition modelling/ABS filament | 250 |
| Conventional manufacturing | Injection moulding/ABS | 1.80 |
| | Machining/1112 screw-machine steel | 0.66 |

modelling, etc.) that need to be removed and constitute waste. The powder-based processes do not tend to require supports. However, the most widely used process for Rapid Manufacture today, SLS of nylon powder, has significant restrictions on the potential to re-use powder. Current practice for RP of SLS nylon parts recommends that used powder is recycled at a ratio of 1:3 with unused powder, resulting in significant waste. Using SLS for Rapid Manufacture will generally impose more stringent requirements in terms of materials used and so the potential to recycle powder will be further reduced. In many cases it is likely that the use of virgin (unused) powder may be a manufacturing requirement in order to maintain product quality and repeatability. The issue of powder re-use for SLS has attracted much interest, resulting in many companies developing re-use strategies. The University of Louisville has developed a method for testing the suitability of used and virgin SLS powders using a bench top melt-flow index test system, allowing mixing in ratios that minimise defects in laser sintered parts [4]. This material characterisation represents a step towards some of the quality assurance issues that will need to be incorporated into the RM systems of the future.

## 9.4 Labour Costs

In general the labour costs associated with Rapid Manufacture are lower than those for machine and material costs. However, this aspect can vary due to the part size and complexity, manufacturing process used, degree of finishing required, production volumes and hourly labour costs.

Labour costs bring up a number of discrepancies between costs for RP and costs for RM. Firstly, setting up a build on an RP machine is a highly skilled and relatively slow process, and so is relatively expensive as part orientation and placement will have a significant effect on the parts produced with various RP processes. This expensive process needs to be repeated for each build with RP. However, the cost is incurred only once for a multitude of builds when series production by Rapid Manufacture is performed. An incidental yet significant benefit of this approach is that machine 'behaviour' can be closely understood and accounted for to achieve improved consistency when using the same machine to produce the same build repeatedly. Even in the Rapid Manufacture of customised products such as hearing aids (see Chapter 13) the geometries being produced are largely similar to each other and part orientation and placement becomes a relatively trivial task.

A second major discrepancy between labour costs for RP and RM lies in part finishing; with RP a skilled operator is required to apply specialist techniques to finish different parts to a high standard and ideally should have a full understanding of the requirements for each particular part:

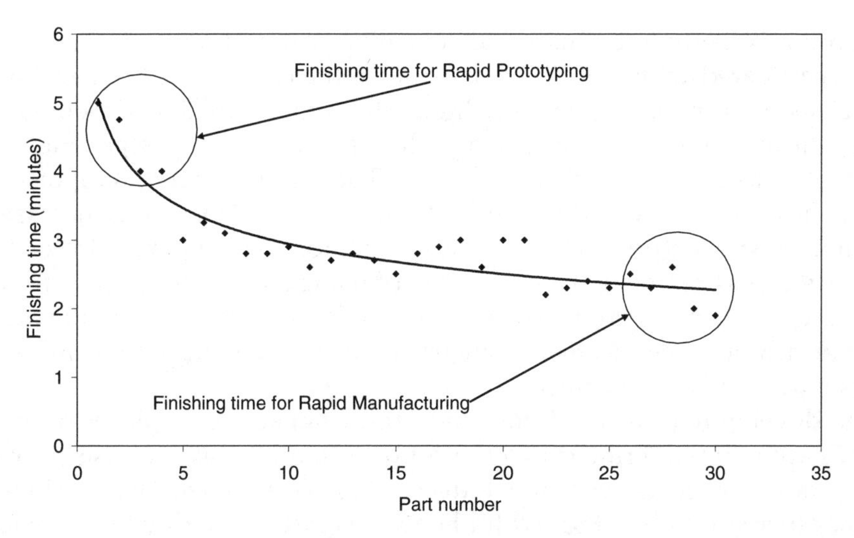


**Figure 9.1** Manual finishing times recorded for a series of small components made by stereolithography

should all stair stepping be removed or are certain tolerances close enough such that hand finishing should be avoided, etc.? In the case of RM, where identical or very similar parts are being produced repeatedly, the range of skills required by the operator are reduced and times required to finish parts will also fall. Figure 9.1 shows the reduction in times required to finish a series of a small component made by stereolithography [5]. The study used to generate the data shown in Figure 9.1 ensured that the quality of part finishing was maintained such that there was no evidence of supports for all parts. Needless to say, a quicker finishing time coupled with a lower skills requirement for RM results in lower finishing costs than for RP. Labour costs may be further reduced if automated finishing can be applied – Align Technologies have developed an automated finishing system for their moulds for dental aligners [6] and this will become a widespread practice for many other RM applications in the future.

Using current RP machines for Rapid Manufacture generally results in a part production cost breakdown within the ranges given below [7]:

- Machine, 50–75 %
- Materials, 20–40 %
- Labour, 5–30 %

These figures vary according to the process and material used, size and complexity of parts made, labour rates, etc. An example of differences between processes may be shown by comparing SLS and fused deposition modelling (FDM) to manufacture an identical geometry. With SLS the weighting is likely to be higher for machine cost and lower for labour cost when compared with FDM. This is because the cost of an SLS machine and associated equipment is generally higher than that for FDM without gaining a significant build speed advantage; however, finishing is generally considerably easier for SLS than for FDM. The use of soluble supports with FDM, however, may reduce the labour cost for FDM but increase the machine costs. A study in 2000 showed that, for a small part with relatively simple support removal, the extra costs of using soluble supports with FDM were greater than the predicted reduction in labour costs they would achieve [7]. For a more complicated geometry with internal supports soluble supports may have been justified.

The development of machines that are designed for Rapid Manufacture (see Chapter 5) will bring down the contribution of machine costs to overall part production costs over time. Indeed Align Technologies have achieved a 3 times increase in build speed [6] hence a significant reduction in machine cost by adapting their stereolithography machines for high-volume manufacture of moulds for dental aligners. Additionally, economies of scale will come into play as RM is more widely adopted – this will bring down costs for both machines and materials. The likely long-term consequence of this is that labour costs will comprise an increasing portion of the production costs of parts made by RM, but automated finishing may reduce this.

## 9.5 Comparing the Costs of Rapid Manufacture with Injection Moulding

The following pages report findings from different studies to consider the economic possibilities of using RP machines for manufacture in place of injection moulding. It must be stressed that the studies were performed on the basis of assuming that the properties of parts produced on RP machines, such as mechanical properties and surface finish, were suitable for end-use products. This assumption is valid in some cases and not valid in others, but for the purposes of a comparison between process purely on production economics it is applicable. It should also be noted that production economics do not take into account the many system economics benefits afforded by RM that are discussed in the following chapter.

In 1997 an initial study by Jonathan Keane and Phill Dickens at Nottingham University looked at the production economics of producing three different components on RP machines and comparing the costs with those of the standard method of manufacture – injection moulding. The premise of this work was to find the 'cut-off' production volume at which RM would become more expensive than injection moulding. Put more simply, if one were to manufacture a single component then RM would be the cheaper option as the cost of injection moulding tooling would be prohibitive.

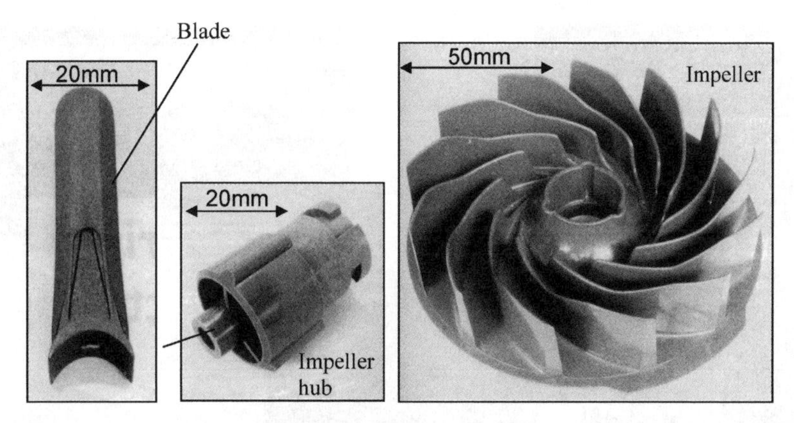


**Figure 9.2** Injection moulded parts selected for the initial cost comparison in 1997

However if 1 million components were required then injection moulding would be the cheaper option as the cost of tooling could be amortised across the large number of parts. The question arose concerning the production volume at which RM becomes more expensive than injection moulding – the so-called 'cut-off volume'.

The components selected for the initial study by Nottingham University were from Flymo, a manufacturer of garden products, and are shown in Figure 9.2 [8]. The results from this study showed the predicted cut in volumes, when comparing injection moulding with SLS on a DTM Sinterstation 2500 (T), to be:

- 6150 for the blade
- 2800 for the impeller hub
- 315 for the impeller

Three years after the original study described above, a similar cost analysis was performed at De Montfort University, UK, along with Delphi Automotive of France [7]. On this occasion four different injection moulded geometries of different sizes but all with complicated geometry were selected (see Figure 9.3).

In this second cost analysis, factors such as energy costs and building space were included, but these only had a minor contribution to overall production costs – less than 1 % in total [7] – and are not included in the discussion below. Costs for injection moulding were obtained by quotes for tooling and unit costs for each part moulded.

Figure 9.4 shows the cost per part versus production volume for the smallest part by injection moulding, fused deposition modelling (FDM2000), stereolithography (SLA7000) and selective laser sintering (EOSP360). As expected the costs for injection moulding are very high for low volumes

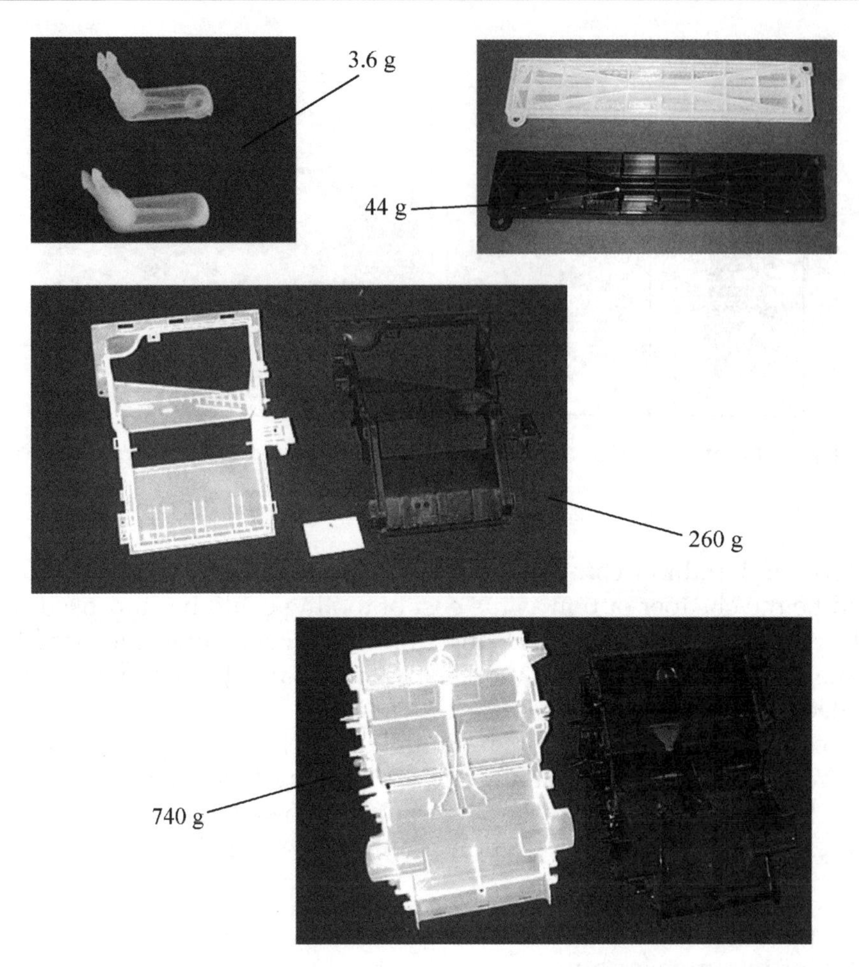


**Figure 9.3** Injection moulded and stereolithography parts used for the second cost analysis project

and reduce as the total cost is amortised over a greater volume of parts. The part costs for SLA7000 and FDM2000 are very similar at approximately 5 euros per part. The costs for selective laser sintering using the EOSP360 machine were significantly lower at 2.20 euros per part – this was largely due to the fact that selective laser sintering allows parts to be stacked on top of each other with 1056 parts in a build, compared with 190 and 75 parts for stereolithography and fused deposition modelling respectively. Also, finishing of the selective laser sintered parts was quicker than that for the stereolithography and fused deposition modelling parts. The fact that SLS proved to be the cheapest production process becomes more significant when one considers that mechanical properties, including stability over time, of such parts are likely to be preferable in terms of required functionality for RM.

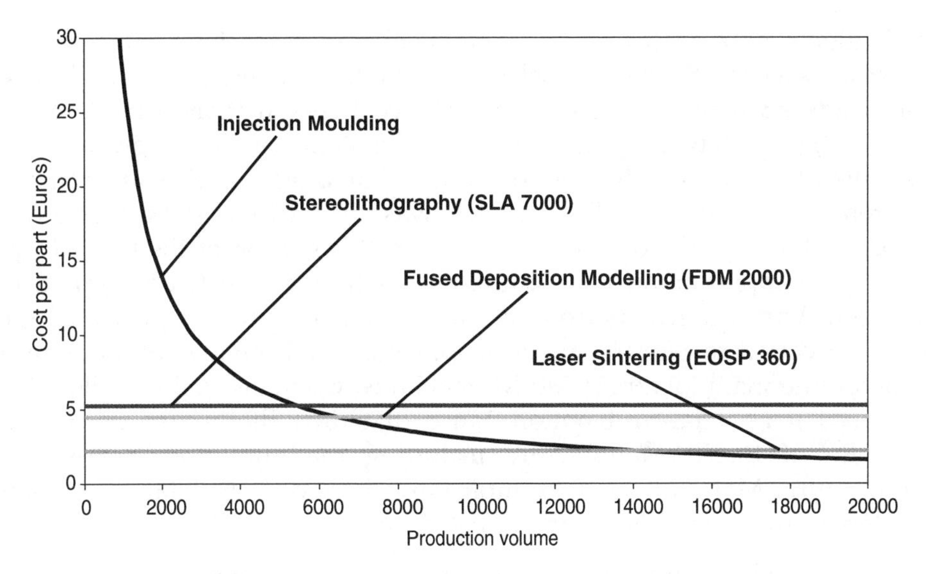


**Figure 9.4** Cut-off volumes for the 3.6 g part by different processes

Table 9.2 shows the cut-off volumes for all four geometries by stereolithography and indicates a heavy reliance on part weight to cut-off volume. However simpler geometries would yield lower cut-off volumes as mould tooling would be cheaper.

The economic production cut-off points found in the studies described above are very much in the realm of batch manufacture, which would normally be more likely to be achieved by machining than by injection moulding. However, in most cases the geometries would not be suitable for machining. This highlights the fact that in any given case the correct solution may not always be clear – for example, could the geometries shown above be re-designed to allow machining? Conversely, could RM be employed to combine numerous components as one part to make RM more competitive?

Another issue to consider is that the studies above were performed on machines that have since been superseded by new, quicker versions. This prompts the question of what would be the price per part for selective laser sintering when performed on a dual laser EOSP700 machine or a 3D Systems Vanguard high-speed machine? An even more compelling case is likely to be found when employing one of the new generation of emerging RP processes (see Chapter 5) for manufacture. As an example EnvisionTec claim that the

**Table 9.2** Cut-off volumes for all four geometries by stereolithography and injection moulding

| Part weight (g) | 3.6 | 44 | 260 | 740 |
|---|---|---|---|---|
| Break-even figure | 5800 | 875 | 336 | 279 |

production cost of a hearing aid on the Perfactory™ machine will cost $0.27. Processes such as SpeedPart, selective inhibition sintering and high-speed sintering (see Chapter 5) are likely to achieve a combination of the low costs given by the Perfactory™ process along with material properties given by SLS.

At the start of this chapter, the issue that costs for RM are radically different from those for RP and RT was introduced and described as probably the most important message from this section of the book. Hopefully the examples above have illuminated why these costs are so radically different. When the results from the studies above were first presented they created a varied response – many people embraced the idea of medium- to high-volume manufacture using RP machines wholeheartedly while others dismissed it as a fanciful notion. Since that time there have been many examples of Rapid Manufacture including medium- and high-volume manufacture. Many of these examples are discussed in detail in the later chapters of this book but a few are listed below:

1. Hearing aids. These are manufactured in quantities of hundreds of thousands by selective laser sintering and stereolithography, where every geometry is unique [9].
2. Electrical connectors. These are manufactured in batches of 16 000 by stereolithography, where every geometry is the same [10].
3. Pill delivery tubes. These are manufactured at a rate of 7000 per year by fused deposition modelling, where every geometry is the same.

When it comes to the choice of whether to adopt RM or not organisations should consider both the production economics as presented in this chapter and the systems economics presented in the next chapter. However, the task of deciding whether to adopt RM or not should be simpler than making the decision as to whether to adopt RP or not. The reason for this is that the costs associated with production of parts by RM (production economics) are simple to calculate and can be weighed against alterative processes and the market value of the products produced. However, in the case of RP the decision to adopt the technology is based almost entirely on the added value of obtaining parts quicker than by other processes – a far less easily quantified figure. Thus, making the case for RM will be far more clear-cut than for RP.

## References

1. Degarmo, E.P., Sullivam, W.G. and Bontadelli, J.A. (1997) *Engineering Economy*, 10th edn, Prentice-Hall, Englewood Cliffs, New Jersey.
2. Wohlers, T. (2003) Rapid Prototyping Tooling and Manufacture Annual State of the Industry Report, Wohlers Associates, USA.

3. http://polymertrack.com (visited 3 June 2004).
4. Gornet, T., Davis, K.R., Starr, T.L. and Richardson, K.M. (2002) Materials characterisation and testing of Duraform PA and GF materials, in Proceedings of the Selective laser sintering User Group Meeting, San Francisco, California, September 2002.
5. Nichols, H. (2003) Rapid manufacturing cost analysis, Final Year Dissertation submitted to Loughborough University, May 2003.
6. Kaza, S. (2002) Utilising stereolithography in mass customisation, in Proceedings of the Stereolithography User Group Meeting, Cost Mesa, California, March 2002.
7. Hopkinson, N. and Dickens, P.M. (2003) Analysis of rapid manufacturing – using layer manufacturing processes for production, *Proceedings of the Institute of Mechanical Engineers, Part C: Journal of Mechanical Engineering Science*, (C1), 31–9, Professional Engineering Publishing Ltd, London.
8. Keane, J.N. (1997) Requirements for rapid manufacturing, Final Year Dissertation submitted to the University of Nottingham, 1997.
9. Caloud, H., Pietrafitta, M. and Masters, M. (2002) Use of selective laser sintering technology in direct manufacturing of hearing aids, in Proceedings of the Selective Laser Sintering User Group Meeting, San Francisco, California, September 2002.
10. Griesbach, S. (2003) Digital manufacturing of connectors in quantities over 500, in Proceedings of the Stereolithography Users Group Meeting, Destin, Florida, March 2003.

# 10

# Management and Implementation of Rapid Manufacturing

Chris Tuck and Richard Hague
*Loughborough University*

## 10.1 Introduction

The advantages of Rapid Manufacturing (RM) lie in the ability to produce highly complex parts that require no tooling and thus a reduction in the costs of manufacture will be possible, especially for low-volume production (Hopkinson and Dickens, 2001; Griffiths, 2002). As high volumes do not need to be manufactured to offset the cost of tooling the possibilities for affordable, highly complex, custom parts becomes apparent. In theory, each product manufactured could be individual.

This chapter aims to provide an overview of how RM could impact the future of manufacturing. The chapter is not concerned with technical details of the RM system or materials, but on the end use of these systems for part and product manufacture.

In order to manipulate RM into a sustainable business proposition, many aspects need consideration. Examples of the areas for consideration are: cost, logistics, supply chain management and issues of change. This chapter aims to give a generalised view of the possible impact RM can have on a manufacturing business. What is without doubt is that the unique attributes

*Rapid Manufacturing: An Industrial Revolution for the Digital Age*
Editors N. Hopkinson, R.J.M. Hague and P.M. Dickens 

of RM will lead to different business paradigms when compared to traditional manufacturing enterprises.

## 10.2 Costs of Manufacture

The costs of manufacture were dealt with more fully in Chapter 9. However, it is important to identify key differences in the costs of the manufacturing system that occur due to the additive manufacturing method.

The difference between RM and traditional manufacturing methods are obvious to those who have been involved with Rapid Prototyping (RP) machinery in the past. One key aspect is that RM does not carry a heavy labour burden and therefore the production of components will require a minimum of operator involvement and for much of the time machines can operate independently. This has an obvious impact on the manner in which a manufacturing environment operates, as systems require no input from operators during part building. Therefore, it is feasible for parts to be built overnight or at weekends when traditional manufacturing companies would need to either halt production or increase workers pay rates at these times. As such, the total number of production hours of the machines is increased without the need to move to shift patterns in order to have skilled machine operators on site. Thus, the majority of costs will result from the machine, its capital cost, energy consumption and usage.

## 10.3 Overhead Allocation

The application of overheads to parts and products will change for additive manufacturing methods. As RM is a technology-intensive process and has a reduced labour requirement, overheads are likely to move from labour content (as is traditional) to the RM machine. Due to the RM process a direct relationship exists between the time necessary for production and the part size or number of parts to be built. Hence, the key variable in production is build time. For this reason, it is more suitable to attribute overheads to the number of build hours available from the machine and thus the longer a part takes to build the more overheads that part consumes.

## 10.4 Business Costs

The ability to get products to market quickly (days/weeks rather than months/years) due to the removal of tooling will mean profits could be taken quicker, allowing a company to lead the market place. As no tooling is involved, risks in new product launches will be reduced as outlays for tooling are eliminated. This will also impact the re-working of existing products, allowing design changes to be brought to market much faster. Significant improvements in operational efficiency, along the lines of lean or agile manufacture, will also be possible. This will be discussed in more detail later in this chapter.

## 10.5 Stock and Work in Progress

In a traditional manufacturing and supply environment, stock can represent both waste (and hence inefficiency) or, conversely, agility (the ability to meet fluctuating demand). For this reason different supply chain management (SCM) paradigms have been introduced in order to eliminate waste or re-distribute stock throughout the supply chain in order to enable flexibility in the manufacturing environment (Walter, Holmström and Tuomi, 2002). The application of RM for suitable parts and components, especially those that are of low volume but high value, can result in a significant reduction in stock costs and inventory levels.

The requirements for stock in a fully Rapid Manufactured product are the computer aided design (CAD) file of the product and raw material for the RM machine to build the item. Designs are held virtually and as such take up nothing but disk space. Raw materials for production can also be used more effectively with an RM system. Raw materials are only manipulated during the period of manufacture and this produces the end result for a complete RM product. Thus, work in progress (WIP) could be eliminated during production; effectively WIP is carried on the RM machine. Thus, any stock held is simply raw material or finished product, removing the need to carry stock of many items for future assembly. Therefore significant reductions in costs of physical stock, WIP and stock holding could be realised, giving enormous benefits to industry.

Reduction in stock can have a large effect on the economics of the business. For example, looking at accounting measures for a business, an immediate effect can be seen on current assets, which have been defined by Reid and Myddleton (1997) as:

> ...all assets that are likely to be soon turned into cash or consumed by the business.

A reduction in stock and WIP, and hence current assets, will have a subsequent knock-on effect on a key accounting measure, working capital (WC). Reid and Myddleton (1997) have defined WC as:

... the 'circulating capital' of the business. It is the excess of current assets over current liabilities.

or

$$\text{Working\_Capital} = \frac{\text{Current\_Assets}}{\text{Current\_Liabilities}}$$

where current liabilities are any creditors due payment within the next year. Although two short-term items define working capital, the balance of WC requires long-term financing. Thus, if WC can be reduced, cash can be released to the business. The reduction in stocks and WIP will immediately have an effect on current assets and thus WC.

Coupled with this reduction in stock, a reduction in logistics may be achievable with parts and products built locally (trackside, for example). This may reduce the need for suppliers and hence impact on current liabilities.

## 10.6 Location and Distribution

The technical requirements of RM systems (raw materials and data) and the relatively small footprint of current RP systems could mean that manufacturing can occur in different environments instead of the traditional factory. Of course the manufacture of components has to occur somewhere, but where? Additionally, what effects will this have on both up- and downstream operations?

A number of different scenarios for production are available for discussion and are only limited by one's imagination. A list (by no means exhaustive) of possible production scenarios is shown below:

- Hub-based manufacture
- Manufacturing on the logistics chain
- Trackside production
- Production at the retailer
- Centralised manufacture and distribution

*Hub manufacture* presents a possibility where designs are purchased and directly sent to a local hub for RM production. The implications for this scenario may be that delivery times are reduced, which brings into question the logistics strategy to exploit remote design and local manufacture. The ability to produce RM parts locally will also have implications on the current globalisation culture of manufacturing business.

*Manufacturing on the logistics chain* has come to prominence as a repair and maintenance scenario for the US Army under the name Mobile Parts

Hospital (MPH). The ability to transport the manufacturing system to remote locations enables the field production of spare parts for military hardware. Considering this in a civilian environment, it may be possible to locate the RM system on trucks or ships for production of components during the delivery phase of the supply chain. Alternately, taking a customisation perspective, it may be possible to customise a product during delivery. For example, a vehicle may be being delivered by ship from overseas; this delivery is effectively dead time on the supply chain where no value is being added to the product. If an RM machine were placed on the ship it could produce components for fitting while at sea and thus shortening times in the factory and adding value to the vehicle.

*Trackside production* is a possible scenario envisaged to take advantages of RM at the manufacturing site. In simple terms, RM will be used to produce parts on demand, sited next to the manufacturing line of a particular product. The key advantages of this scenario are that a reduction in logistics and WIP during manufacture could result.

*Production at the retailer* is another scenario based on providing an RM function at the retailer of a particular product. For example, RM could be based at local automotive dealerships for the production of modular upgrades or customised components to correspond with the customer's requirements. This has further advantages for the automotive manufacturer; the production of a base platform (i.e. a car with basic engine and trim options) can be achieved and time-consuming customer options fitment can be postponed until the product is at the dealer. Therefore, the basic configuration can be produced at lower costs through standardised manufacturing methods at the manufacturing site.

*Centralised manufacture and distribution* has been investigated by Walter, Holmström and Tuomi (2002). They have considered the affect RM can play on distribution and logistics for the aerospace spare parts business. The supply chain for some aerospace components was too slow and demand too unpredictable to continue with existing supply chain practices. Therefore RM methods were sought and evaluated. Two scenarios were considered: firstly, the centralised production of components to replace inventory and, secondly, a decentralised set of RM units to replace inventory and some of the conventional distribution network (a hub-based system). Taking the first premise, benefits could be realised in the cost of holding stock for slow-moving items that would otherwise be subsidised from more popular, faster selling items. The hub-based scenario was thought to be suitable only when demand was high enough at a given location.

These scenarios are but a few of those conceivable, but importantly they all have RM at the core. A common consideration for these scenarios is that the manufacturing operation can be tailored to the product. Taking some of the examples shown above, further advantages from RM can be seen.

## 10.7 Supply Chain Management

The effect that RM has on the production environment, location of manufacture and distribution of components all require managing as in a traditional manufacturing business. The notion of supply chain management (SCM) has been developed to manage the supply chains involved in business effectively. The supply chain is formed by the network of suppliers that are involved in providing products or services to a firm. The supply chain includes raw materials suppliers, subassembly, information systems, logistics, retailers and finally the customer. SCM is concerned with the management of the supply chain as a whole (Slack *et al.*, 1998). A key objective for businesses developing and researching their supply chain management practices is to:

> . . . appropriate value for themselves from their participation in the supply chain (Cox, 1999a).

The ability of RM to appropriate value from SCM practices stems from its flexibility of production. RM will impact different SCM methods in differing ways. It should be noted that not all supply chain methods and practices are suitable for a given company. Demarcation in supply chain practice should be developed by analysing a firm's business practices and the properties of demand for its products. Fisher (1997) has developed a generic framework for identifying the correct supply chain focus for different products. This is shown in Figure 10.1.

Figure 10.1 shows areas where SCM methods match the product. In essence for a functional product (where demand is predictable) a lean supply chain is most suitable, whereas for an innovative product (where demand is unpredictable) an agile supply chain is preferred. Development

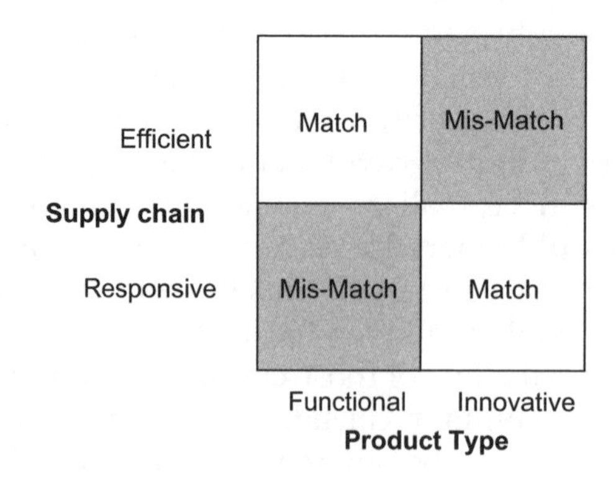


**Figure 10.1** Product, supply chain matrix (Source: Fisher, 1997)

of a suitable supply chain practice for the appropriation of value has resulted in a number of supply chain concepts. These include:

- Lean
- Agile
- Leagility and postponement
- Mass customisation
- Demand chain

It is not the intention of this chapter to discuss the finer details of these paradigms, but to acknowledge their importance in modern-day manufacturing and discuss the possible impact of RM upon them.

### *10.7.1 Lean*

The lean manufacturing paradigm relies on eight fundamental principles for successful implementation, as discussed in Cox (1999b), that provide the backbone to the lean methodology. RM will affect the 'mechanics' of the lean methodology and therefore the production of parts and the elimination of waste in the supply chain.

#### Produce Only When Necessary, Through Just-in-Time

The application of RM techniques will hold a number of advantages for just-in-time (JIT) manufacturing, which include:

1. *Dematerialised supply chain.* The overriding requirement for RM is to have suitable three-dimensional CAD data from which to produce the part or product. This will have consequences upon the supply chain as it could be said that the supply chain has been dematerialised.
2. *True just-in-time.* As the RM machine requires only three-dimensional CAD data and raw materials to produce a part, the application of RM could result in a reduction of material distribution and stock holding or warehousing costs for WIP. The ability to amalgamate RM with Internet technology and other manufacturing systems (materials resource planning, or MRP, etc.) could lead to JIT manufacture *at* the manufacturing site, rather than the traditional concept of JIT delivery *to* the manufacturing site.
3. *Reduced set-up, changeover time and number of assemblies.* It must be stressed that the production of parts through RM will change the manufacturing paradigm from that of skilled labour operating machinery and forming a large portion of the part cost to one where the burden of cost is transferred to the technology or specifically the RM machine and materials. A further driver for the reduction of costs may be in the product design. For example, RM processes may make traditional designs obsolete

by reducing the need for assemblies and thus the production process, providing cost savings for parts and components (Hague, Campbell and Dickens, 2003).

As RM requires no tooling changes to produce different parts or products, time that is traditionally lost to these factors will be reduced and hence a second element of the lean paradigm will be addressed – that of the elimination of waste.

### Elimination of Waste

A principle driver for the lean paradigm is to reduce waste in the supply chain. Considering the digital supply chain (Fralix, 2003) and integration with Internet technology will result in the exchange of data between designers and manufacturers. With RM these design data could be sent directly to an RM system for building. The production of parts to order quickly and economically should result in no stock holding unless this is desired for reasons of agility (Fisher, 1997).

By enabling the digital supply chain and RM in the same manufacturing system, current manufacturing systems will be turned on their heads. Today, manufacturing tends to occur in centralised factory environments. However, taking into account the scenarios presented earlier, RM could be located locally, even in the home. Thus the design of components could be done remotely with subsequent local manufacture.

It may be possible to take the concept of RM in the lean supply chain further. If production is reliable and does not result in stock penalties, can production be brought back in house? For example, production of parts and components could occur next to an assembly line. What implications would this have on the supply chain? Primarily this will have a drastic effect on the cost of logistics. It may no longer be necessary to ship goods, either internationally or nationally. This will affect the costs of production and thus allow supply chains practising lean methodologies to further reduce waste.

The factors discussed above will result in the elimination of waste in terms of:

- Material
- Time
- Costs
- Distribution

The impact on agile or responsive supply chains will be no less profound and those effects seen on lean supply chains will also impinge on the agile paradigm.

### *10.7.2 Agile*

Naylor, Naim and Berry (1999) have defined the term agility in his discussion of lean, agile and leagile supply chains:

> Agility means using market knowledge and a virtual corporation to exploit profitable opportunities in a volatile market place.

Agile supply chains work on a different set of operational methodologies. Agility focuses on lead-time compression, rather than the elimination of waste. The use of flexible production methods allows fast reconfiguration of processes to cope with consumer demand. It is the variability of consumer demand that defines the motive for the agile manufacturing paradigm. For these reasons the agile paradigm is suited to products that have a short lifecycle, such as fashionable goods, compared with lean production's focus on commodity production. The market order winner for agile supply chains is no longer cost but availability (Mason-Jones, Naylor and Towill, 2000). The driver for successful implementation of an agile supply chain would therefore be market information, the forecasting of demand. Thus the information regarding a customer's preferences drive the pull of products from the manufacturing environment, with an associated increase in cost to accommodate the inevitable increase in costs associated with changes in build methods and products.

It could be said that the advent of the agile supply chain has been necessary because of the increased demands of the customer being placed on the producer (Maskell, 2001). The sophistication of customer desires and tastes has led to goods becoming increasingly fashion oriented, with styles and colours becoming a market order winner rather than product function. One methodology for enabling this 'value-adding' activity has been the concept of mass customisation. The ability to define a customer's needs and wants relies on the availability of suitable information on customer preferences and the ability within the agile organisation to provide these services in a timely manner. Thus, the emphasis on production has changed from one based on costs of production to knowledge and information availability, and hence the skills and knowledge of the organisation are now paramount (Maskell, 2001).

### *10.7.3 Leagility and Postponement*

Agile supply chains are designed to cope with volatile products, such as those that are fashionable or have a short product lifecycle. The advent of RM would mean that lean production in a responsive manner could become a reality without the need for a 'leagile' concept (the combination of lean and agile supply chains) (Mason-Jones, Naylor and Towill, 2001).

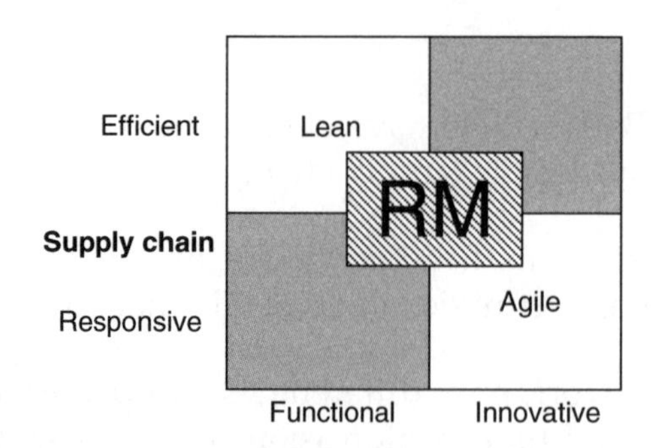


**Figure 10.2** Modified production and supply chain matrix

In effect the manufacturing machine would become a de-coupling (or postponement) point, where orders are only produced at the customer's request. With this concept, there would be no stock-outs, as all products could be made to order, plus the threat of obsolescent stock would be negated as the only stock necessary to hold would be raw material and design data. Taking Fisher's $2 \times 2$ matrix as seen in Figure 10.1 with modifications to take into account the application of RM technology, the effects are obvious (Figure 10.2).

The current mis-match that exists with an efficient supply chain and an innovative product can be removed using RM techniques, thus providing an innovative product with a lean and efficient supply chain mechanism. In summary, RM could offer:

- A truly 'leagile' supply chain
- Low-cost products with fast re-configurability and fast response
- Reduction in stock levels
- Reduced waste
- Increased value through customisation
- Reduced logistics cost
- Reduction in part count
- Increased flexibility

### *10.7.4 Impact of RM on Mass Customisation*

Mass customisation (Pine II, 1993) is a method that uses standard products as a base for customisation of certain characteristics of the product, adding value to that product. RM could have an important role in core customisation, a concept described by Alford, Sackett and Nelder (2000), where the customer has intimate involvement with the production and design of a vehicle. The automobile industry offers numerous options on their products from the colour of bodywork to type of seat fabric, as well as the finance deal that the customer is offered. These customisation issues are not considered as core customisation and in an article by Fox (2003) it was suggested that core customisation of cars was not yet possible. However, taking into account the advantages of RM, this may now be possible (Loch *et al.*, 2003). The ability to produce complex structures from digital data taken from or designed with the customer could be manufactured and used on that customer's vehicle. RM could offer true core customisation on two levels:

1. *Aesthetic customisation.* The first level of customisation would be that based on customer preference, which may be to improve the aesthetics of parts or components to the customer's specification.
2. *Body-fit customisation.* The second level involves the customer more intimately. By digitally capturing the customer's body shape it may be possible to provide the customer with body-fit parts, e.g. a customised seat and seat back, to provide a more comfortable and possibly safer environment.

In order to manufacture these types of products traditionally the process of design and production would involve a great deal of skill and hence a large expense. With the advent of RM, the production method and processes involved for customised parts would not change from part to part. Thus, the economic argument for providing this core customisation method would be improved. The design and customer input issues of customisation have been discussed in Chapters 2 and 3 respectively.

### *10.7.5 RM and the Demand Chain*

The demand chain has been studied by Frohlich and Westbrook (2002), where the management of the demand chain has been defined as:

> ... practice that manages and coordinates the supply chain from the end-customers backwards to the suppliers.

The practice of demand chain management (DCM) exists to provide transparency of information when considering the demand for products. Thus, a principal concern is the logistics necessary to satisfy demand (Williams, Maull and Ellis, 2002). Internet technologies are seen as a significant enabler for DCM; therefore the combination of RM and Internet technologies needs consideration.

#### **Internet and e-commerce**

RM and the Internet offer advantages over traditional supply chain concepts. As mentioned previously, current fundamental requirements for RM machines are power, raw material and data. Taking this into consideration, data and data delivery will become critical issues for concern. Recent studies of Internet systems saw large advantages when applied to inventory management, with some organisations able to reduce their stock holding by around 4 % (Lancioni, Smith and Schau, 2003).

The introduction of Internet technology will affect the application of RM in a number of ways. The effects could be noticeable on inventory costs, data management, planning and procurement, distribution and manufacturing strategy. The greatest effect on all of these is to increase the amount of flexibility afforded to the organisation. The reduced necessity to hold finished stock and the ability to transfer data efficiently will result in reduced costs both in terms of waste and time.

### Data Management

The ability to manage key data, such as the CAD files necessary for manufacture, will be a key driver in the implementation of RM technology. The Internet should become a key enabler for this as long as data transfer rates and data integrity can be guaranteed. The amalgamation of the Internet and RM will result in questions of how to best link the two. Work by Rajagopalan *et al.* (1998) has shown the coupling of manufacturing techniques with Internet technology to form a virtual manufacturing infrastructure. This infrastructure comprised three main elements: a design client, a process broker and the manufacturing service.

In simple terms, a design is produced and a suitable production process chosen by the broker who then distributes information to the manufacturer. The design is presented to the broker as a CAD file; the broker then manipulates the file to form a suitable medium for the manufacturer. The interfaces were facilitated by plug-in software distributed to the designers and manufacturers to interact with the broker. While this infrastructure was a success and enabled the production of components designed remotely from the manufacturer, efficiency could be compromised by the presence of the broker. The advent of Internet technologies and appropriate software should mean that designers could integrate directly with manufacturers.

The transparency afforded by direct contact through the Internet and speedy data transfer suggests that planning and procurement activities could be streamlined. For example, the necessary part is designed and a manufacturer sought with sufficient capacity or suitable lead-time. Using Internet technology, real-time data should be available and the manufacturer could be chosen.

## 10.8 Change

All of the discussions so far in this chapter have pointed to areas of business where RM could have an effect, whether this is reducing stock or producing customised goods. What is clear is that RM will change manufacturing organisations. Change will occur both on a technical standing, where re-training and familiarisation will be necessary, and also on a managerial level.

The ability to introduce this new technology coherently and efficiently will be necessary when RM systems come online, but RM will also provoke change at the organisational level. For customised or personalised products, data collection and design methods will need to be re-thought. Depending on the supply chain scenario chosen, many decisions will need to be made on whether to purchase RM equipment and manufacture in house or purchase RM components from a supplier, thus impacting logistics and distribution. Possibly the largest but unknown impact could be on company culture and how it changes to accommodate RM.

## 10.9 Conclusions

In conclusion, this chapter has attempted to outline some of the effects on the business, its operation and principles that may occur with the advent of RM. This will lead to the economic production of single units and the economic production of low to medium volume products, as demonstrated by the literature. The use of RM will enable the world's first truly flexible manufacturing unit to produce a part of almost any geometry and complexity while being able to produce a single unit of a desired product economically.

RM has the ability to modify manufacturing operations greatly. The opportunities available for reduction in costs of production, through the natural rationalisation of logistics, labour, stock holding and the ability to deal with unstable demand patterns, are all apparent. The ability to remove these costs could also affect the manufacturing environment on a global scale, by returning manufacturing to the country of origin, as labour costs are no longer a burden. RM could realise the first truly flexible and JIT supply chain paradigm to respond to customer demand and changes in taste and design with ease.

Additionally, the introduction of RM could lead to increased value in products through the realisation of truly customised production. Though many questions remain to be answered on how the development of RM and full customisation can be implemented, the impact of this technology on the manufacturing environment will be profound. The impact for the mass customisation community will be that those products which one may not have been able to produce, either because the market was too small (one to a few thousand units) or the design was far too complex, will now be possible. Put simply, no market will be too small, no part too complex and manufacturing will be able to offer a complete customer solution on time and at reasonable cost.

## References

1. Alford, D., Sackett, P. and Nelder, G. (2000) Mass customisation – an automotive perspective, *International Journal of Production Economics*, **65**, 99–110.
2. Cox, A. (1999a) A research agenda for supply chain and business management thinking, *Supply Chain Management: An International Journal*, **4**(4), 201–11.
3. Cox, A. (1999b) Power, value and supply chain management, *Supply Chain Management: An International Journal*, **4**(4), 167–75.
4. Fisher, M. (1997) What is the right supply chain for your product?, *Harvard Business Review*, **75**(2), 105–16.
5. Fox, S. (2003) Recognising materials power, *Manufacturing Engineer*, April.
6. Fralix, M. (2003) The realities behind the myths of mass customisation, Mass Customisation and Personalisation Conference, Munich, October 2003.
7. Frohlich, M.T. and Westbrook, R. (2002) Demand chain management in manufacturing and services: web-based integration, drivers and performance, *Journal of Operations Management*, **20**(6), 729–45.
8. Griffiths, A. (2002) Rapid manufacturing – the next industrial revolution, *Materials World*, **10**(12), 34–5.
9. Hague, R., Campbell, I. and Dickens, P. (2003) Implications on design of rapid manufacturing, *Proceedings of the Institution of Mechanical Engineers, Part C: Journal of Mechanical Engineering Science*, **217**(C1), 25–30.
10. Hopkinson, N. and Dickens, P. (2001) Rapid prototyping for direct manufacture, *Rapid Prototyping Journal*, **7**(4), 197–202.
11. Lancioni, R.A., Smith, M.F. and Schau, H.J (2003) Strategic Internet application trends in supply chain management, *Industrial Marketing Management*, **32**, 211–7.
12. Loch, C., Sommer, S., Schafer, G. and Nellassen, D. (2003) Will rapid manufacturing bring us the customised car?, *Automotive World Knowledge*, March 2003, www.acknowledge.com.
13. Maskell, B. (2001) The age of agile manufacturing, *Supply Chain Management: An International Journal*, **6**(1), 5–11.
14. Mason-Jones, R., Naylor, B. and Towill, D.R. (2000) Engineering the leagile supply chain, *International Journal of Agile Management Systems*, **2**(1), 54–61.

15. Naylor, J.B., Naim, M.M. and Berry, D. (1999) Leagility: integrating the lean and agile manufacturing paradigms in the total supply chain, *International Journal of Production Economics*, **62**, 107–18.
16. Pine II, J. (1993) *Mass Customisation – The New Frontier in Business Competition*, Harvard Business School Press, Boston, Massachusetts.
17. Rajagopalan, S., Pinilla, J.M., Losleben, P., Tian, Q. and Gupta, S.K. (1998) Integrated design and rapid manufacturing over the Internet, in Proceedings of ASME Design Engineering Technical Conference 1998 (*DETC98*), Atlanta, Georgia.
18. Reid, W. and Myddleton, D.R. (1997) *The Meaning of Company Accounts*, 6th edn, Gower Publishing Limited, p. 7.
19. Slack, N., Chambers, S., Harland, C., Harrison, A. and Johnston, R. (1995) *Operations Management*, Pitman.
20. Walter, M., Holmström, J. and Tuomi, J. (2002) Rapid manufacturing and its impact on supply chain management; http://www.tai.hut.fi/ecomlog/publications/rapid_manufacturing.pdf.
21. Williams, T., Maull, R. and Ellis, B. (2002) Demand chain management theory: constraints and development from global aerospace supply webs, *Journal of Operations Management*, **20**, 691–706.

# 11

# Medical Applications

Russ Harris and Monica Savalani
*Loughborough University*

## 11.1 Introduction

One of the greatest benefits of Rapid Manufacturing (RM) is that it enables the economically viable production of an article in which all the design features are in direct response to a given application. There are minimised compromises imposed on the design by the limits of the production process or the necessity to produce goods that are of a 'best fit' such as with mass production. Such capabilities will allow the viable manufacture of one-off articles that are custom-made in respect of shape and functionality. Each of our bodies are different, thus providing a vast arena of applications for custom-made products.

Medical applications and research in RM are driven by an individual's unique requirements of shape and functionality, and the value of impact. There are also large financial incentives. An indication of the size of the market is well illustrated by the UK's increasing expenditure on public healthcare, provided in Figure 11.1.

An indication of the value of RM to the healthcare industry is demonstrated by the financial investment in a European research project called Custom-Fit. Beginning in 2004 for a period of 5 years and a total budget of 18 million ECU, Custom-Fit concerns the research and exploitation of RM with a particular emphasis on medical applications [2].

This chapter presents some of the activities of RM in the field of medical treatment and healthcare. Such applications of new technologies are much longer term than other engineering fields due to the stringent and

*Rapid Manufacturing: An Industrial Revolution for the Digital Age*
Editors N. Hopkinson, R.J.M. Hague and P.M. Dickens 

| Period | Expenditure (£ millions) |
|---|---|
| 1991/92 | 31,842 |
| 1992/93 | 35,413 |
| 1993/94 | 37,259 |
| 1994/95 | 39,879 |
| 1995/96 | 40,691 |
| 1996/97 | 42,383 |
| 1997/98 | 43,878 |
| 1998/99 | 47,194 |
| 1999/00 | 48,362 |
| 2000/01 | 53,039 |
| 2001/02 | 59,852 |

**Figure 11.1** UK Government expenditure on the National Health Service (Office for National Statistics)(1)

multi-faceted requirements for clinical approval. Thus, the applications described in this chapter are at various stages of exploitation. Some are relatively well established, e.g. surgical planning, while others should be considered as ongoing research topics, e.g. *in vivo* implantation devices.

## 11.2 Pre-Surgery RM

Some of the earliest medical applications of layer manufacturing and associated technologies have been exploited in pre-surgical activities. Whether such applications fit within the strict definition of RM is open to discussion. In several applications the object manufactured is directly utilised to produce a desired 'end product', i.e. surgical simulation. However, this may be viewed as part of a process and is not the end product, which is an effective medical treatment. However, the benefits and acceptance that have been realised through these works have instigated continued research and further applications of RM technologies in healthcare and therefore justify their discussion at the beginning of this chapter.

High-complexity surgical procedures can have a duration in excess of 10 hours. The longer this period, the greater the risk to the patient. Thus, any planning, practice, simulation, evaluation and decision making that can be performed pre-surgery is of great value. Surgical planning tools represent some of the earliest examples of RM in healthcare applications.

Applications are normally dependent on X-ray computer tomograph (CT) or magnetic resonance imaging (MRI) patient data and include:

- Pre-operative planning
- Pre-forming of fixation components
- Manufacture of surgical guides and templates
- Simulation of surgical procedures

- Fit evaluation of implants
- Patient demonstration
- Intraoperative guidance
- Surgeon training and tangible recording

Various research groups and surgeons have evaluated the use of RM in surgical planning. The commercial manufacture of these products is particularly specialised by two major companies – Materialise and Anatomics.

Some examples of the value of pre-surgery RM techniques are summarised below. Minns *et al.* [3] reported the use of RM to produce a pre-operative solid stereolithography (SL) geometry from a patient with a defect of the medial tibial plateau (the top of the tibia) of the knee for proposed reconstructive surgery. The geometry gave the surgeon both the three-dimensional anatomical information needed to ascertain whether there was adequate bony support after cutting to fit a prosthesis, as well as a solid product on which to simulate the proposed surgery, before undertaking the procedure on the patient.

Schenker [4] reported the usefulness of reverse engineering and RM technologies in the emergence of new tools in medicine. The possibility of viewing and physically handling the precise geometry before surgery provided great benefits in evaluation of the procedure and implant fit in difficult cases. The use of RM provided lessened risk to the patient and reduced cost through reduced theatre time. A case study was presented involving hip replacement in a patient who had experienced severe bone loss through osteoporosis.

Anatomics have previously presented a case concerning the use of RM in pre-surgical planning in the separation of Siamese (conjoined) twins in Brisbane. A series of SL BioModels™ from Anatomics were used by an interdisciplinary team of neurosurgeons, plastic surgeons and reconstructive surgeons to assist in the surgical planning. The BioModels™ highlighted the conjoined vein network of interest in the babies (illustrated in Figure 11.2) and were built from a file created by Anatomics BioBuild™ software using a high-resolution CT scan with intravenous contrast enhancement of the blood vessels. The detailed full-scale BioModels™ allowed the many surgeons

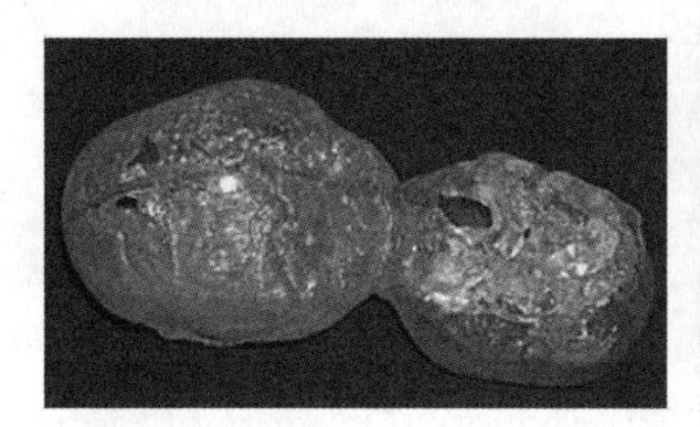
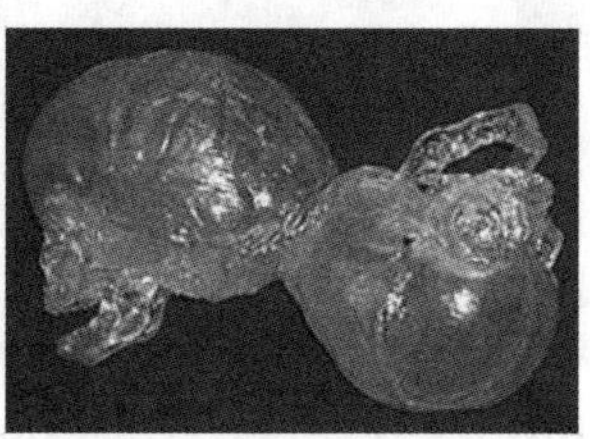

**Figure 11.2** RM BioModels™ for complex surgical planning. (Reproduced by permission of Anatomics Pty Ltd)

involved to assess and plan the optimal separation accurately and to study the relationship of crucial blood vessels in close proximity to the skull and each other. The BioModels™ were constructed using selectively colourable StereoCol resin, to enhance visualisation of the blood vessels with red coloration. The solid replicas were a critical physical communication tool for the surgical team in planning and during the complex 12 hour operation. The use of the BioModels™ was identified as being crucial to the success of the procedure and the reduction of complications. Both babies were eventually moved from post-operative intensive care sooner than expected and the separation operation was considered a complete success.

An example of RM enabling precise implant production is provided by Materialise and University Hospital of Rotterdam who presented the case of a 24-year-old male with a large craniofacial traumatic frontal bone defect. Both protective and cosmetic reasons made reconstruction of this defect necessary. A polymethylmethacrylate (PMMA) cranioplasty implant was indirectly produced via a SL model of the damaged area. Subsequently, precise fitment of the implant shortened the operation time and eliminated the need for any corrections. The natural contour of the skull was re-established and no complications were encountered. The process cycle is illustrated in Figure 11.3.

Anatomics have presented a single case where RM has been implemented in surgical planning, mirroring and implant production (illustrated in Figure 11.4). The case concerned a 60-year-old female who presented a large right frontal bony growth (hyperostotic meningioma) which was removed. The patient subsequently exhibited a reduced orbital volume and distinct cosmetic deformity. The proposed subtractive surgery was initially simulated utilising BioModels of the patient's cranial geometry. Post-sugery, another BioModel™ was generated from which an implant was modelled, produced and evaluated. The surgeon estimated that at least 1 hour of operating time was saved in both surgical phases. The time saving

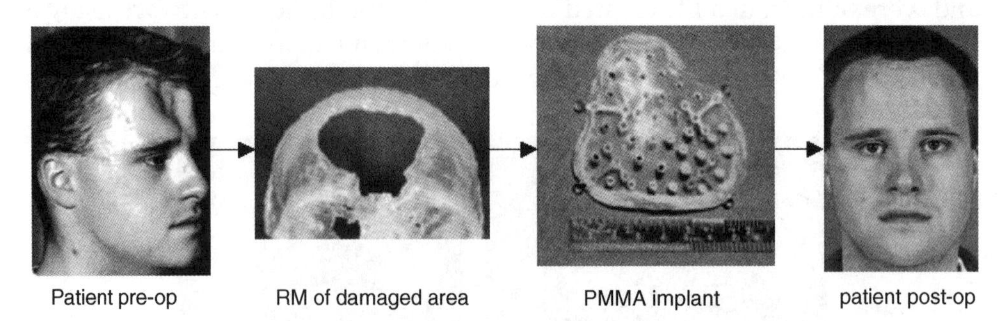


**Figure 11.3** Craniplasty implant via RM visualisation. (Courtesy of Materialise NV and Dr J.K.Th. Haex, H.W.C. Bijvoet and A.H.G. Dallenda, University Hospital of Rotterdam, Dijkzigt)

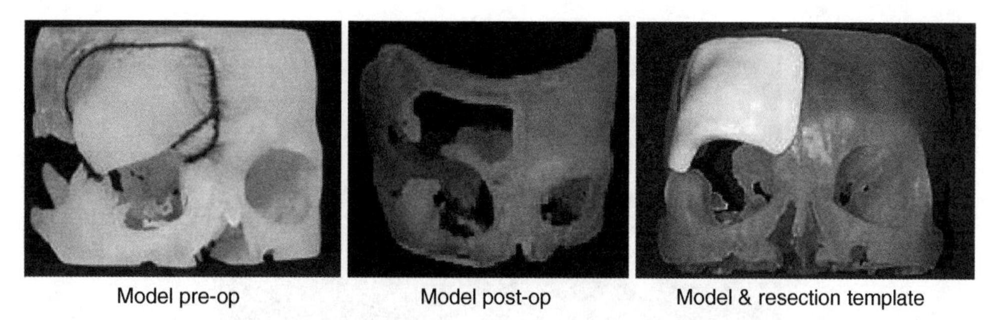


**Figure 11.4** Surgical planning, mirroring and implant fitment (Reproduced by permission of Anatomics Pty Ltd)

was due to the surgeon confidently implementing the tumour resection on the basis of pre-operative planning and by the use of the prefabricated implant.

The use of RM for constraint-free investigation and risk-free surgical simulation highlights its usefulness as an education tool. Suzuki *et al.* [5] reported the role of RM in the education of medical students with respect to ear surgery. This surgery is especially delicate and the bone structures exceptionally small. It has been previously demonstrated that the skills of ear surgery are best developed by dissecting a temporal bone (the skull structure found around the ear). However, only a limited number of trainees can be afforded this opportunity because of the scarcity of available bones. In order to address this, models were built by RM (selective laser sintering (SLS), glass-filled polyamide) technology. These were similar in hardness to real bone, with an accurate reproduction of surface structures. This could be shaved using a surgical drill, burr and suction irrigator in the same way as a real bone. The malleus and incus (constituting parts of the temporal bone structure) were reproduced, along with the semicircular canals and the oval and round window niches. It was concluded that such RM models serve as a good educational tool for middle ear surgery.

Prior works have demonstrated the many applications and value of RM in surgical planning. It can be seen that in such applications RM is facilitating increased treatment speed and efficiency. The trends exhibited reveal that the use of RM in surgical planning is moving beyond its initial applications as a visualisation tool for surgeons and into an engineering tool for implant production and reconstructive treatments.

## 11.3 Orthodontics

The use of RM has been heavily utilised in orthodontics, in particular for oral implantation. Artificial tooth implants are attached to the mandible and

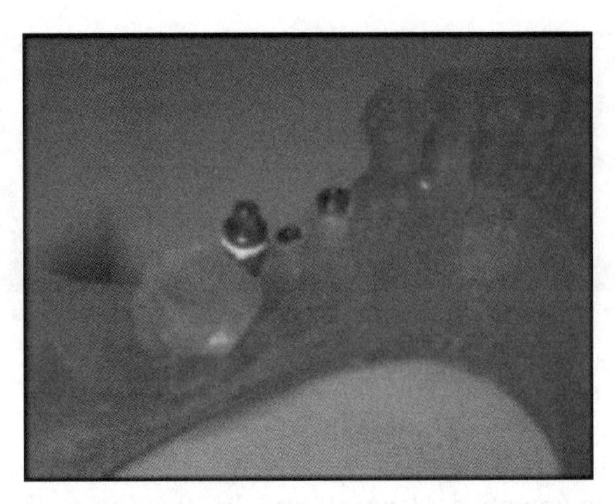

**Figure 11.5** Implant simulation. (Reproduced with permission of AITeM, Associazione Italiana di Technologia Meccanica from Campanelli, L.A.C. De Filippis, A.D. Ludovico and A. Falco, stereolithography to the service of dental implantology, in Proceedings of the 6th AITeM International conference, Cassino-Gaeta, Italy, 8–10 September 2003)

maxilla by screw fixation. The drilling of relatively long holes into such narrow bone structures at a precise position requires great accuracy and when conducted manually is prone to error. Reverse engineering and RM have been utilised to allow the production of assisting tools and guides that provide improved accuracy, greater treatment speed and virtually zero risk of misplacement.

SL models have been used by Campanelli *et al.* [6] in pre-operative planning and surgery simulation for dental implants (demonstrated in Figure 11.5). The transparency of the model provided the surgeon with information of the internal structure, which made it easier to work. After verification it was clear that the correct positioning of the implant makes it a very powerful tool for pre-operative simulation and verification.

Orthodontic drill guides have been shown to relieve surgeons of concerns regarding the positioning and placement which enabled more efficient treatment [7]. This study compared the accuracy of conventional surgical guides to that of an SL surgical guide and reported an improvement in implant placement with the help of the SL surgical guide.

The company Materialise produce software for determining and generating the necessary guide geometry derived from inspection data (i.e. CT), and also provide a service of manufacturing the guides. These are called SimPlant and SurgiGuides respectively. Examples are shown in Figure 11.6.

Research has been conducted into the RM of the dental implants themselves. Selective laser melting (SLM) has been investigated by Kruth *et al.* [8] for the production of titanium dental prosthetic frameworks. The use of SLM

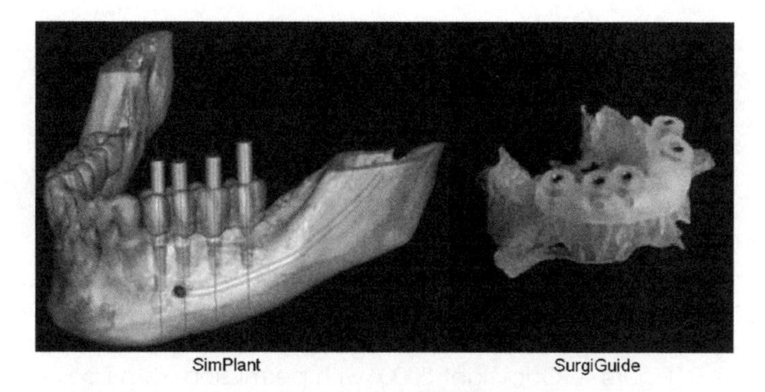


**Figure 11.6** Example of implant planning with SimPlant and guide template by SurgiGuides. (Reproduced with permission of Materialise NV)

was investigated in preference to selective laser sintering (SLS) with the aim of producing fully dense parts in one step, thus avoiding the delicate cleaning procedure of the green part and the time-consuming oven infiltration cycle. The work reported a current inaccuracy of the frameworks originating from the 'stair-stepping' effect common to many layer-wise production techniques.

The successful exploitation of RM technologies in high volumes for orthodontic applications is represented by Align Technologies, Inc., who produce the Invisalign® system for teeth straightening. The treatment consists of the patient wearing a sequential series of customised clear plastic aligners. Each aligner in the series is produced indirectly from individual SL models. Align Technologies claim 40 000 treatments to date with 18–30 aligners (and therefore SL models) produced for each case.

The use of RM in orthodontics has been an application that has been quickly commercially exploited. The technology has made previously delicate and demanding implantation procedures much more feasible and subsequently is becoming increasingly commonplace.

## 11.4 Drug Delivery Devices

Oral drug delivery pharmaceuticals are constructed using multiple components. Akin to this, many RM methods operate by assembling and joining multiple components together to produce a solid article (e.g. SLS and three-dimensional printing both utilise powdered base materials). Subsequently, there have been several investigations into the use of RM techniques for controlled production of complex, multi-component oral drug delivery tablets.

Researchers have exploited the capabilities of three-dimensional printing (3DP) for RM of polymeric drug delivery systems to overcome the drawback

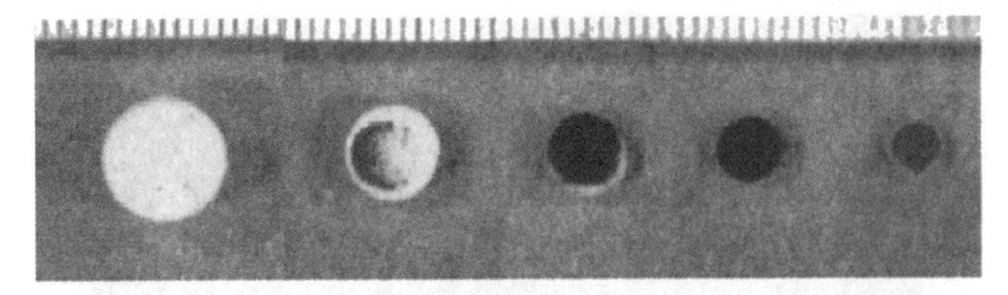

**Figure 11.7** 3DP tablets subjected to dissolution testing at various times. The scale bar at the top of the photograph is in millimetres. (Reprinted from *Journal of Controlled Release*, Vol. 66, W.E. Katstra *et al.*, oral dosage forms fabricated by Three Dimensional Printing$^{TM}$, pp.1–9, Figure 3, Copyright 2000, with permission from Elsevier)

of the decreased drug release rate as a function of time. A drug concentration profile can be generated in a computer model which may be produced by RM (also see Chapters 4 and 7 on CAD and FGMs respectively). By this means complex drug delivery regimes can be created featuring multiple drugs or the periodical, dosed release of a single drug. Wu *et al.* [9] emphasised that dosage changes can be achieved by altering the printing parameters during component construction. The study demonstrated three means of controlling the delivery regime: the three-dimensional position, microstructure and composition. Katstra *et al.* [10] reported the use of 3DP for the fabrication of controlled dosage delivery pharmaceuticals with complex internal geometries, varying densities and diffusivities. Subsequently, erosion mechanisms for delayed-release tablets were constructed by varying the composition (see Figure 11.7). Further research [11] reported the fabrication of four types of multi-mechanism oral drug delivery devices using 3DP. The tablets fabricated were classified as immediate-extended release, breakaway (a fast eroding interior releasing two extended drug releasing subunits), enteric dual pulsatory release (two releases in the intestine) and dual pulsatory release (two releases, one in gastric fluid and one in the intestine). Alteration of tablet properties were achieved by variation in operating parameters, including composition variation, layer thickness, line spacing, flow rate and axis binder speed.

Liew *et al.* [12, 13] reported the capability, concepts and possibilities of building oral drug delivery pharmaceuticals using SLS. It was identified that current polymeric drug delivery devices lack precision. As a result, the efficiency and effectiveness of drug delivery is impaired. The first aspect of the study focused on a 'space' creation technique by varying densities by SLS process alteration. Results indicated that varying laser power during sintering enabled the creation of channels in which a second material could be deposited. The further development of polymeric-controlled drug delivery devices using the SLS process depended heavily on the ability to incorporate a bioactive agent (secondary material) within a suitable biocompatible polymer (primary material). In the second aspect of the study, emphasis was placed on a secondary powder deposition method to selectively deposit the powder-based material based on electrography. By incorporating the secondary depository system and using the space creation technique, the results indicated some success for the fabrication of customised polymeric matrix-controlled drug delivery devices with dual materials.

The use of RM for controlled drug delivery devices is a topic that could have large areas of healthcare application. One single tablet with controlled release of different drugs could replace the necessity to take several oral drugs over a period. Also, there is the possibility for customised drug delivery where the drug components, dosage and delivery lags are tailored to an individual patient's requirements.

## 11.5 Limb Prosthesis

As mentioned previously, each of our bodies vary in their geometry and so create many demands for custom-made products. Also, when our bodies are damaged, such as limbs being lost, the damaged area is likely to be of a unique geometry. In the case of prosthetic limbs these damaged areas become interfaces between the body and the prosthetic. The fit at this interface is especially critical due to the need for accurate mechanical load transfer and comfort. The need for customised products is further emphasised by the frequent revisions required due to the changing geometry of the damaged limb. There have been several examples of research efforts in RM for limb prosthesis.

In 1998 Freeman and Wontorcik [14] evaluated the use of SL in the direct manufacture of prosthetic limb interfaces/sockets for fit and comfort testing. At this time the SL materials available were too brittle to withstand long-term use. Also, the manufacturing costs were shown to be higher than conventional methods.

Later research demonstrates the progress of RM technologies for limb prosthesis. Ng, Lee and Goh [15] reported the development of a dedicated RM system for the direct production of polypropylene (PP) prosthetic interfaces/sockets manufactured using a polymer deposition technique in which a socket is formed by a continuous strand of partially melted PP that is spirally deposited according to the socket's cross-sectional contour. The working principle of the system was very similar to fused deposition modelling (FDM) but with a lower level of precision and accuracy, which was sufficient in the application and provided faster build speeds (greater extrusion widths than FDM); both of these factors allowed more economical production as compared to FDM. The system reduced the socket-making time from days for traditional processes to less than 4 hours by the RM process. Preliminary investigation revealed that the functional characteristics were similar to that of a traditional socket. Goh, Lee and Ng [16] further investigated the material strength (ASTM D638-99) and structural integrity (ISO 10328) of these PP sockets with particular concern regarding potential delamination. No problems with structural integrity were found while a 13–23 % lower ultimate tensile strength (UTS) as compared to conventional PP sockets was overcome by using a double, wall arrangement in the region where failure occurred.

The rapid production of customised prosthesis interfaces provides the recipients with an efficient form of treatment that is tailored to their specific and changing requirements in respect to comfort and functionality. Although, in relation to some applications, the demand for such goods is relatively low, the production of these interfaces demonstrates how RM can be simply applied to the benefit of a medical/healthcare sector and individuals.

## 11.6 Specific Advances in Computer Aided Design (CAD)

The use of RM in many areas of application, including medicine and healthcare, is dependent upon synchronous developments in associated design software. To date, the increasing use of RM for medical applications has been accompanied by advances in related design software, such as the previously mentioned SimPlant software by Materialise.

McGurk *et al.* [17] reported that rapid advances in computer technology had created new possibilities in surgery which previous generations of surgeons could only have imagined. Improvements in computerised tomography (CT) and magnetic resonance imaging (MRI) have provided the data from which physical objects may be realised by RM.

Vander Sloten, Van Audekercke and Van der Perre [18] reported advances in computer aided engineering (CAE) software that have set new standards in the design of prostheses. The design of implants has benefited by the ease of obtaining an internal digital geometric description (via CT, etc.) that has allowed the design derivation and fit evaluation of newly designed implants at an early stage. Device size and design optimisation may be assisted by numerical evaluation of implant strength, stability of fixation, etc. Advances in information technology have also enabled the transfer of digital patient images into CAD software, providing fast and even semi-automatic implant design. RM technology can then be used for cost-effective production of this, one of a kind, product.

In addition to the dental drill guides discussed in section 11.3, work has been conducted to produce SL drill guides/templates for accurate location of Branemark implants (titanium screws that are driven into bone

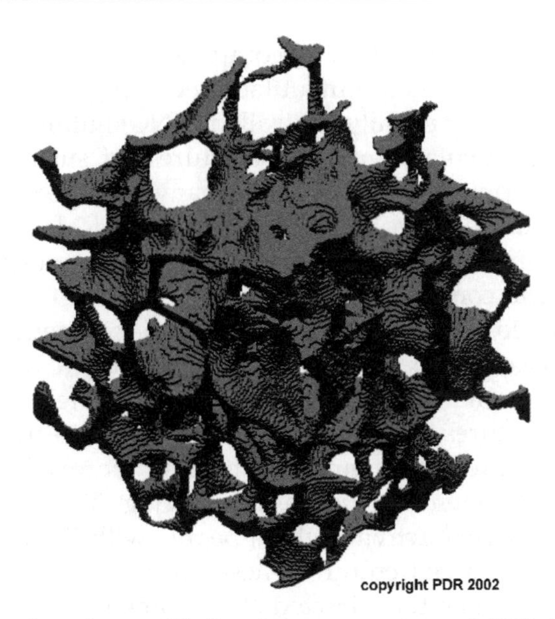


**Figure 11.8** Cancellous bone .STL file. (Image courtesy of PDR, University of Wales Institute Cardiff)

and pass through the skin) that will hold prosthetic ears. This work by Bibb *et al.* [19] fully exploited the advantages of CAD by constructing the necessary template design in the virtual environment without physical models. The resultant design was generated by SL and directly utilised in surgery.

Bibb *et al.* have also highlighted how the design capabilities of RM may be restricted by software issues. Some of these issues were encountered in the production of SL representations of cancellous bone structures (the inner bone structures) which may be used for laboratory testing in preference to real bone structures (illustrated in Figure 11.8). This work demonstrated difficulties in the RM of these products due to the increasingly restrictive nature of preparatory software. Although such software is becoming more user-friendly it is much more difficult to alter parameters for complex and atypical designs.

## 11.7 *In Vivo* Devices

*In vivo* (inside a living organism) devices are artificial devices put into the human body or living organisms. Examples include scaffolds, pins, plates, rods, screws, valves, sutures, grafts and fixations which are manufactured to suit a variety of medical situations. The principle medical requirements of *in vivo* devices include sterility, biocompatibility, biostability and sustainability. Most *in vivo* devices are manufactured using conventional methods including Computer numerically controlled (CNC) milling, casting, forming, etc. Complexity (particulary internal structures) of such goods are highly valued for *in vivo* applications. There are many examples of research being conducted with the aim of producing such goods by RM. The driving force is the potential for greater control of structures and compositions with fewer design and material constraints, greater speed and customisation of *in vivo* devices as compared to conventional fabrication techniques.

For *in vivo* devices there are two main areas of RM research: relatively solid products (i.e. solid bone implants) and deliberately porous structures (i.e. scaffolds – structures that enable and encourage directed tissue growth). The production of extremely complex scaffold structures is a particularly hot topic due to developments in tissue engineering. Tissue engineering is a largely laboratory, based activity which begins with living tissue cells that are multiplied through cell culture. These are then seeded into a carrier/scaffold which facilitates the directed three-dimensional growth and proliferation. This form of regenerative medicine provides replacement tissue and organs and is widely foreseen as one of the next great breakthroughs in medical treatment. Subsequently, there is significant competition to formulate and exploit a suitable three-dimensional containment structure to facilitate the growth of tissue-engineered products.

This section broadly categorises *in vivo* RM research efforts according to their processing nature.

### 11.7.1 Fused Deposition Modelling (FDM) for In Vivo Devices

Kalita *et al.* [20] investigated the development and fabrication of controlled porosity structures with polypropylene (PP) polymer and tricalcium phosphate (TCP) ceramic composite using FDM. Samples with 36 % volume porosity and average pore size of 160 μm showed the best compressive strength of 12.7/MPa. The results showed that samples were non-toxic and also supported excellent cell growth during the first two weeks of *in vitro* (simulation of conditions inside a living organism) studies.

Hutmacher *et al.* [21] reported the fabrication of bone scaffolds for complex craniofacial skeletal reconstruction utilising medical imaging, computational modelling and FDM. The study indicated that the use of defect-specific polycaprolactone (PCL) scaffolds produced by FDM were promising alternatives to conventionally produced scaffolds.

Zein *et al.* [22] also investigated the effect of process variation in the fabrication of bioresorbable PCL porous scaffolds using FDM. Alteration of processing parameters allowed the fabrication of scaffolds with a varying channel size and porosity. Results showed that compressive properties had a high correlation with the degree of porosity, regardless of the lay-down pattern and channel size used in scaffold fabrication.

Too *et al.* [23] investigated the development and feasibility of producing accurately controlled three-dimensional non-random porous structures (i.e. scaffolds) with FDM. They reported the effect of FDM process parameters (road width, slice thickness, raster gap setting) on important scaffold properties (porosity, pore diameter, compressive strength). Micrographs indicated that three-dimensional interconnecting pores existed within the microstructure of the specimens. It was noted that raster gap settings had a major influence on the microstructure of the specimen.

### *11.7.2 SLA (Stereolithography Apparatus) for In Vivo Devices*

Presently no stereolithography (SL) materials have been clinically approved for *in vivo* devices. Medical grade materials are available that can be sterlised and temporarily exposed to human fluids but these cannot be left inside the body after surgery. Subsequently the *in vivo* applications of SL are currently restricted to indirect production techniques.

Sodian *et al.* [24] reported the indirect fabrication of tissue scaffolds for heart valves using SL. The SL models were then used to generate biodegradable and thermoplastic elastomer polymer scaffolds (poly-4-hydroxybutyrate and polyhydroxyoctanoate) via a thermal processing technique.

Levy *et al.* [25] have produced biodegradable scaffolds indirectly with SL. The intended model is produced by SL using hydroxyapatite (a calcium-based ceramic that naturally occurs in bone) powder suspended in the photosensitive resin. The SL polymer is then burnt out, leaving a sintered porous hydroxyapatite structure.

### *11.7.3 SLS for In Vivo Devices*

Pioneering work in the early 1990s by Lee and Barlow [26] investigated using selective laser sintering (SLS) for producing bioceramic *in vivo* implantation devices. They combined acrylic polymer latex and polymethylmethacrylate (PMMA) with hydroxyapatite for SLS, with the polymer being burnt out in a secondary process. The work highlighted the interconnected porosity, which would be of value for bone ingrowth and later tissue engineering scaffolds.

Tan *et al.* [27] demonstrated the processing of polyetheretherketone (PEEK) and hydroxyapatite (HA) powder blends by SLS. The laser power, part bed temperature and scan speed were the three SLS parameters investigated. Results indicated the potential production of PEEK/HA scaffolds by SLS, with 40 % HA content ideal for biological purposes and optimum SLS process parameters of a lower part bed temperature complemented by a higher laser power.

Das *et al.* [28] investigated the SLS of polyamide nylon (PA6) scaffolds with varying biomimetic internal architecture, constructed using computational design methods. Although the studies were largely orientated towards assessing the scaffold design, the final PA6 scaffolds showed significant potential for *in vivo* use. Biocompatibility testing supported cell viability quite well and this success would instigate further studies of multi-component graded architectures.

Chua *et al.* [29], reported the production of viable scaffold designs by SLS with consistent and reproducible microarchitectures, which were designed by utilising a parametric library and assembly algorithm. The scaffolds were not built with a biomaterial but with commercial SLS nylon (Duraform™) since this particular research focused largely on assessing the design possibilities. The varied parameters used on the Sinterstation 2500 included: power, fill scan speed, powder layer thickness, warm-up height, cool-down height and part bed temperatures. However, removing the loose powder within the pores proved to be difficult.

Chua *et al.* [30] later reported the fabrication of poly(vinyl alcohol) (PVA)/HA composites by SLS to fabricate tissue engineering scaffolds. Promising specimens were produced with optimum SLS parameters of a power bed temperature of 65 °C, scan speed of 1270 $mms^{-1}$ with a laser power of 15 W. Bioactivity analysis showed compatibility of the specimens within a simulated human body fluid environment.

### 11.7.4 3DP for In Vivo Devices

Lam *et al.* [31] investigated the production of scaffolds using three-dimensional printing (3DP)/plotting with a starch-based polymer. A blend of starch, based polymer powders suitable for 3DP (50 % cornstarch, 30 % dextran and 20 % gelatine) was developed. The structures were post-processed by infiltrating with a co-polymer to increase strength and water absorption resistance. The scaffolds were examined for their microstructure, porosity, mechanical properties and water absorption levels. The results showed that it was possible to produce scaffolds with these materials using 3DP, but its biocompatibility was not determined.

The Bioplotter, invented at the Freiburg Materials Research Centre and produced by Envisiontec GmbH, is a dedicated, commercially available RM machine which operates under cell compatible conditions and may be used in the production of scaffolds incorporating living cells (illustrated in Figure 11.9). This technique has been used by Landers *et al.* [32] who reported the building of scaffolds using hydrogels (a polymer network with a high water content) intended for the engineering of soft tissue structures. This work produced hydrogel scaffolds with a specific external shape and a defined internal pore structure. The geometry of a nose

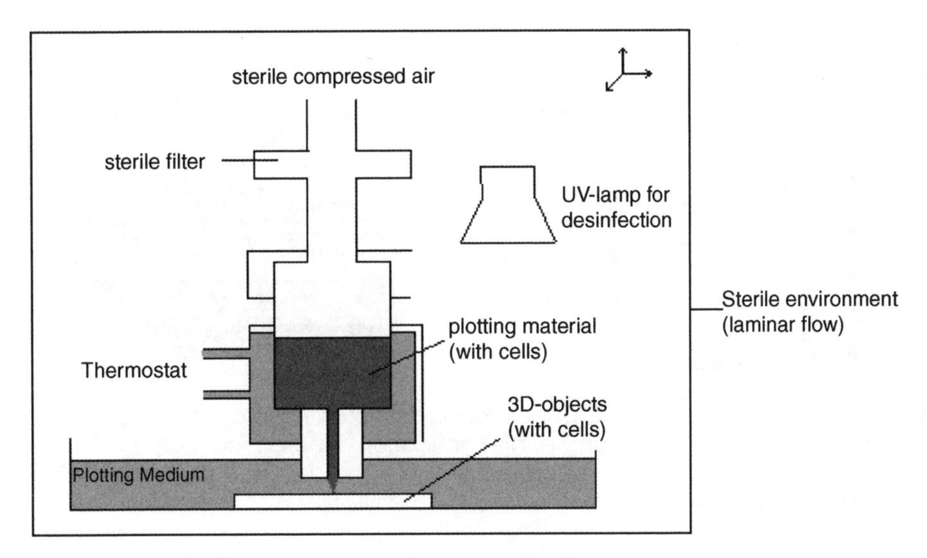


**Figure 11.9** Bioplotter schematic. (Reprinted from *Trends in Bio-technology*, Vol. 21, No. 4, V. Mironov *et al.*, organ printing: computer-aided jet-based 3D tissue engineering, pp. 157–61, Figure 4, Copyright 2003, with permission from Elsevier)

was fabricated to demonstrate the potential for soft tissue implant generation.

Mironov *et al.* [33] discussed the concept, practical aspects and developments for RM cell printing for direct soft tissue engineering. The process discussed involved three sequential steps: computer aided design of the organ (pre-processing), organ printing (processing), organ conditioning and accelerated organ maturation (post-processing). It identified vascularization (the development of vessels) as the main obstacle in efficient organ printing but that computer aided layer-by-layer assembly of biological tissues and organs was currently feasible.

### *11.7.5 Other RM Processes for In Vivo Devices*

Xiong *et al.* [34] reported the fabrication of composite scaffolds for tissue engineering using low-temperature deposition manufacturing (LDM) (illustrated in Figure 11.10). A composite of poly(L-lactic acid) and TCP was evaluated for use in bone regeneration scaffolds. LDM is an RM process based on computer-controlled deposition, which is conducted below 0 °C. By operating without the use of elevated temperatures (such as with SLA, SLS, FDM, etc.) the process aims to better preserve the bioactivity of scaffold materials. The frozen scaffolds produced by the system are subsequently freeze dried to provide a solid, state scaffold at normal atmospheric temperatures. The scaffolds exhibited mechanical properties close to human spongy bone (porous bone structures), but much lower than compact

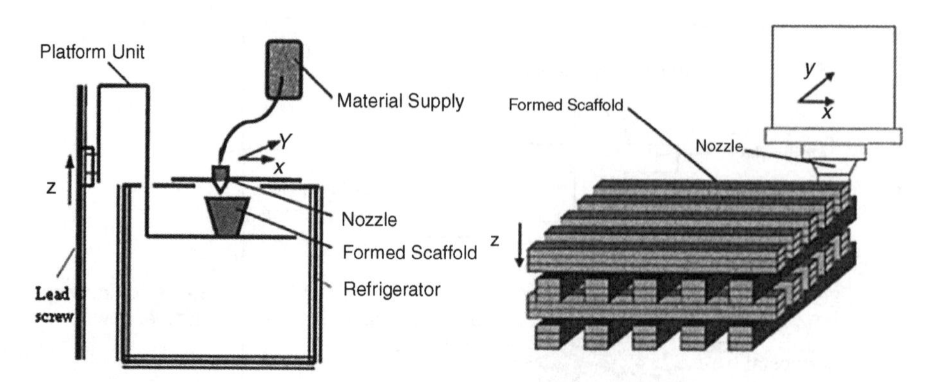


**Figure 11.10** Low-temperature deposition manufacturing process. (Reprinted from *Scripta Materialia*, Vol. 45, Z. Xiong *et al.*, Fabrication of porou scaffolds for bone tissue engineering via low-temperature deposition, pp. 771–6, Figure 1, Copyright 2002, with permission from Elsevier)

bone (dense bone structures). Canine implantation showed good biocompatibility and bone conductivity (the adhesion of existing bone on and into the implant).

Ciardelli *et al.* [35] investigated and compared the fabrication of poly (ε-caprolactone) (PCL) and poly(ε-caprolactone)-poly(oxyethylene)-poly (ε-caprolactone) (PCL-POE-PCL) scaffolds using the pressure-assisted microsyringe (PAM) RM technique and SLS. The results showed that the two techniques were capable of very different structure resolutuions. PAM allowed the fabrication of fine-featured microstructures down to 5–10 μm wide and 5 μm high, whereas SLS structures exhibited a resolution approximately 700 μm wide and 300 μm high. Cell attachment to both scaffolds was very good, which may have been aided by the surface topography and roughness.

Ang *et al.* [36] reported the fabrication of three-dimensional chitosan (a naturally occurring amino-polysaccharide) and hydroxyapatite scaffolds with regular and reproducible macropore architecture using an RM dispensing system. The RM technique was an extrusion based system consisting of a computer-guided desktop robot and a single-component pneumatic dispenser. *In vitro* cell culture studies indicated that the scaffolds were capable of incorporating cell seeding and proliferation.

Woodfield *et al.* [37] investigated the use of a three-dimensional fibre deposition RM technique to produce scaffolds for engineering articular cartilage using molten poly(ethylene glycol)-terephthalate-poly(butylenes terephthalate) (PEGT/PBT). The technique enabled the production of desired scaffold characteristics by controlling the deposition of the molten co-polymer fibres from a computer-controlled pressure-driven syringe. By varying PEGT/ PBT composition, porosity and pore geometry, three-dimensional-deposited scaffolds were produced with similar mechanical properties to native articular cartilage. Cell culture *in vitro* (outside of the body) and subcutaneous (under the skin) implantation in mice showed that the seeded scaffolds supported cell distribution and cartilage-like tissue formation.

As testified by the number of research activities, the potential value of RM for products for use within the body is extremely high. Through different approaches, these activities aim to enable the efficient and automised production of tailor made products for the rapid treatment of patients who have suffered the loss of bone and tissues structures. Unfortunately, such conditions are not uncommon and their typical causes are traumatic injury and destructive surgical treatments. The realisation of RM for such devices will benefit healthcare providers and industry, and, most importantly, the individuals who are experiencing such conditions.

## References

1. Office for National Statistics, *1991/1992–2001/2002: Annual Abstract of Statistics.*
2. Harris, R.A. (2004) *Medical Modeling* (eds A. Christensen and T. Wohlers), Wohlers Report 2004 – Rapid Prototyping and Tooling State of the Industry Annual Worldwide Progress Report, Wohlers Associates, Inc., USA, Part 8: *Other Developments*, pp. 195–206.
3. Minns, R.J., Bibb, R., Banks, R. and Sutton, R.A. (2003). The use of a reconstructed three-dimensional solid model from CT to aid the surgical management of a total knee arthroplasty: a case study, *Medical Engineering and Physics* **25**(6), 523–6.
4. Schenker, R. (1999) Novel combination of reverse engineering and rapid prototyping in medicine, *South African Journal of Science*, **95**(8), 327–8.
5. Suzuki, M., Ogawa, Y., Kawano, A., Hagiwara, A., Yamaguchi, H. and Ono, H (2004) Rapid prototyping of temporal bone for surgical training and medical education, *Acta Otolaryngology*, **124**, 400–2.
6. Campanelli, S., De Filippis, L.A.C., Ludovico, A.D. and Falco, P.A. (2003) Stereolithography to the service of dental implantology, in Proceedings of the 6th AITeM International Conference, Cassino-Gaeta, Italy, 8–10 September 2003.
7. Sarment, D.P. (2003) Accuracy of implant placement with a stereolithographic guide, *International Journal of Oral and Maxillofacial Implants*, **18**(4), 571–7.
8. Kruth, J.P., Van Vaerenbergh, J., Mercelis, P., Savalani, M., Lauwers, B., and Naert, I. (2003) Selective laser sintering of dental prostheses, in Proceedings of Euro-uRapid 2003 International user's conference on

*Rapid Prototyping and Rapid Tooling and Rapid Manufacturing*, Frankfurt, Germany, 1–2 December 2003.

9. Wu, B.M., Borland, S.W., Giordano, R.A., Cima, L.G., Sachs, E.M. and Cima, M.J. (1996) Solid free-form fabrication of drug delivery devices, *Journal of Controlled Release*, **40**(1–2), 77–87.
10. Katstra, W.E., Palazzol, R.D., Rowe, C.W., Giritlioglu, B., Teung, P. and Cima, M.J. (2000) Oral dosage forms fabricated by Three Dimensional Printing™, *Journal of Controlled Release*, **66**(1), 1–9.
11. Rowe, C.W., Katstra, W.E., Palazzolo, R.D., Giritlioglu, B., Teung, P. and Cima, M.J. (2000) Multimechanism oral dosage forms fabricated by Three Dimensional Printing™, *Journal of Controlled Release*, **66**(1), 11–17.
12. Liew, C.L., Leong, K.F., Chua, C.K. and Du, Z. (2001) Dual material rapid prototyping techniques for the development of biomedical devices. Part 1: space creation, *The International Journal of Advanced Manufacturing Technology*, **18**(10), 717–3.
13. Liew, C.L., Leong, K.F., Chua, C.K. and Du, Z. (2002) Dual material rapid prototyping techniques for the development of biomedical devices. Part 2: secondary powder deposition, *The International Journal of Advanced Manufacturing Technology*, **19**(9), 679–7.
14. Freeman, D. and Wontorcik, L. (1998) Stereolithography and prosthetic test socket manufacture: a cost/benefit analysis, *American Academy of Orthotists and Prosthetists*, **10**(1), 17–20.
15. Ng, P., Lee, P.S.V. and Goh, J.C.H. (2002) Prosthetic sockets fabrication using rapid prototyping technology, *Rapid Prototyping Journal*, **8**(1), 53–9.
16. Goh, J.C.H., Lee, P.V.S. and Ng, P. (2002) Structural integrity of polypropylene prosthetic sockets manufactured using the polymer deposition technique, *Proceedings of the Institution of Mechanical Engineers, Part H: Journal of Engineering in Medicine*, **216**(6), 359–68.
17. McGurk, M., Amis, A.A., Potamianos, P. and Goodger, N.M. (1997) Rapid prototyping techniques for anatomical modelling in medicine, *Annals of the Royal College of Surgeons of England*, **79**(3), 169–74.
18. Vander Sloten, J., Van Audekercke, R. and Van der Perre, G. (2000) Computer aided design of prostheses, *Industrial Ceramics*, **20**(2), 109–112.
19. Bibb, R., *et al.* (2003) Planning osseointegrated implant sites using computer aided design and rapid prototyping, *The Journal of Maxillofacial Prosthetics and Technology*, 6, 1–4.
20. Kalita, S.J., Bose, S., Hosick, H.L. and Bandyopadhyay, A. (2003) Development of controlled porosity polymer–ceramic composite scaffolds via fused deposition modelling, *Materials Science and Engineering C: Biomimetic and Supramolecular Systems*, **23**(5), 611–20.
21. Hutmacher, D.W., Rohner, D., Yeow, V., Lee, S.T., Brentwood, A. and Schantz, J.-T. (2002) Craniofacial bone tissue engineering using medical imaging, computational modeling, rapid prototyping, bioresorbable

scaffolds and bone marrow aspirates, in *Polymer Based Systems on Tissue Engineering*, Aloor, Algarve, Portugal, 2002.
22. Zein, I., Hutmacher, D.W., Tan, K.C. and Teoh, S.H. (2002) Fused deposition modeling of novel scaffold architectures for tissue engineering applications, *Biomaterials*, **23**(4), 1169-85.
23. Too, M.H., Leong, K.F., Chua, C.K., Dee, Z.H., Yang, S.F., Cheah, C.M. and Ho, S.L. (2002) Investigation of 3D non-random porous structures by fused deposition modelling, *The International Journal of Advanced Manufacturing Technology*, **19**(3), 217–23.
24. Sodian, R., Loebe, M., Hein, A., Martin, D.P., Hoerotrup, S.P., Potapov, E.V., Hausmann, H.A., Lueth, T. and Hetzer, R. (2002) Application of stereolithography for scaffold fabrication for tissue engineered heart valves, *American Society for Artificial Internal Organs Journal*, **48**(1), 12–16.
25. Levy, R.A., Chu, T.M.G., Halloran, J.W., Feinberg, S.E. and Hollister, S.J. (1997) CT-generated porous hydroxyapatite orbital floor prosthesis as a prototype bioimplant, *American Journal of Neuroradiology*, **18**(8), 1522–5.
26. Lee, G. and Barlow, J.W. (1993) Selective laser sintering of bioceramic materials for implants, in Solid Freeform Fabrication Symposium, University of Texas, Austin, Texas, 9–11 August 1993, pp. 376–80.
27. Tan, K.H., Chua, C.K., Leong, K.F., Cheah, C.M., Cheang, P., Abu Bakar, M.S. and Cha, S.W. (2003) Scaffold development using selective laser sintering of polyetheretherketone–hydroxyapatite biocomposite blends, *Biomaterials*, **24**(18), 3115–23.
28. Das, S., Hollister, S.J., Flanagan, C., Adecounmi, A., Bark, K., Chen, C., Ramawamy, K., Rose, D. and Widjaja, E. (2003) Freeform fabrication of Nylon-6 tissue engineering scaffolds, *Rapid Prototyping Journal*, **9**(1), 43–9.
29. Chua, C.K., Leong, K.F., Tan, K.H., Wiria, F.E. and Cheah, C.M. (2004) Development of tissue scaffolds using selective laser sintering of polyvinyl alcohol/hydroxyapatite biocomposite for craniofacial and joint defects, *Journal of Materials Science – Materials in Medicine*, **15**(10), 1113–21.
30. Chua, C.K., Leong, K.F., Wiria, F.E., Tan, K.C. and Chandrasekara, M. (2004) Fabrication of poly(vinyl alcohol)/hydroxyapatite biocomposites utilizing rapid prototyping for implementation in tissue engineering, in International Conference on *Competitive Manufacturing – Progress in Innovative Manufacturing*, University of Stellenbosch, South Africa, 4–6 February 2004, pp. 229–34.
31. Lam, C.X.F., Mo, X.M., Teoh, S.H. and Hutmacher, D.W. (2002) Scaffold development using 3D printing with a starch-based polymer, *Materials Science and Engineering C: Biomimetic and Supramolecular Systems*, **20**(1–2), 49–56.

32. Landers, R., Pfister, A., Hubner, U., John, H., Schmelzeisen, R. and Mulhaupt, R. (2002) Fabrication of soft tissue engineering scaffolds by means of rapid prototyping techniques, *Journal of Materials Science*, **37**(15), 3107–16.
33. Mironov, V., Boland, T., Trusk, T., Forgacs, G. and Markwold, R.R. (2003) Organ printing: computer-aided jet-based 3D tissue engineering, *Trends in Biotechnology*, **21**(4), 157–61.
34. Xiong, Z., Yan, Y.N., Wang, S.G., Zhang, R.J. and Zhang, C. (2002) Fabrication of porous scaffolds for bone tissue engineering via low-temperature deposition, *Scripta Materialia*, **45**(11), 771–6.
35. Ciardelli, G., Chiono, V., Cristallini, C., Barbani, N., Ahluwalia, A., Vozzi, G., Preoiti, A., Tantussi, G. and Giustl, P. (2004) Innovative tissue engineering structures through advanced manufacturing technologies, *Journal of Materials Science: Materials in Medicine*, **15**(4), 305–10.
36. Ang, T.H., Sultana, F.S.A., Hutmacher, D.W., Wong, Y.S., Fuh, J.Y.H., Mo, X.M., Loh, H.T., Burdet, E. and Teoh, S.H. (2002) Fabrication of 3D chitosan–hydroxyapatite scaffolds using a robotic dispensing system, *Materials Science and Engineering C: Biomimetic and Supramolecular Systems*, **20**(1–2), 35–42.
37. Woodfield, T.B.F., Malda, J., De Wijn, J., Peters, F., Riesle, J. and Van Blitterswijk, C.A. (2004) Design of porous scaffolds for cartilage tissue engineering using a three-dimensional fiber-deposition technique, *Biomaterials*, **25**(18), 4149–61.

# 12

# Rapid Manufacturing in the Hearing Industry

Martin Masters, Therese Velde and Fred McBagonluri
*Siemens Hearing Instruments*

## 12.1 The Hearing Industry

Today's hearing aid manufacturing industry operates in a highly competitive global market place. Hearing instruments are dispensed to users from a wide variety of outlets. There are very small operations with one or two principal dispensers and there are large chains of stores that dispense thousands of hearing aids per year. Governments also dispense hearing aids, e.g. the Veterans Administration in the US and the National Health Service in Great Britain. These dispensers of hearing aids have a large variety of options available to them when it comes to prescribing an instrument.

A modern hearing instrument is a small, powerful electronic signal processor coupled with a microphone and loudspeaker which are used to correct a user's hearing loss. There are two main varieties of hearing instruments available today: behind the ear (BTE) and in the ear (ITE). While both types require the electronics package to be small, the ITE instruments present by far the larger challenge in terms of manufacturing the device. While there is a possible role for mass customization in the BTE devices, most such customization work in the industry to date has been done on the ITE side.

A hearing instrument must fit the user's ear tightly, yet comfortably. This is because instruments tend to go into feedback if there is not a good acoustic seal, especially in cases where significant gain is required to correct for the

*Rapid Manufacturing: An Industrial Revolution for the Digital Age*
Editors N. Hopkinson, R.J.M. Hague and P.M. Dickens 

hearing loss. The situation is analogous to the experience of a person giving a presentation at a convention who causes a loud squeal by walking in front of the loudspeakers with a microphone. The difference is that in the case of hearing instruments the loud squeal happens inside someone's ear. This effect can be painful and embarrassing and therefore does not result in a good listening experience. Such a precise fit is even more difficult to achieve when one considers that the requirement is to fit inside the shell: a microphone, fully programmable digital signal processor, speaker, battery, directional microphones, assorted switches, telephone detector circuitry and volume control knobs!

## 12.2 Manual Manufacturing

Hearing aid manufacturing has evolved over the past few decades but it remained essentially an 'art' during this same period. To replicate the ear of a hearing aid user, a process is used that creates a hollow shell from a series of impressions and moulds. First, impression material that is designed to set rapidly is introduced into the patient's ear. After the material cures or hardens, it is removed, thus creating a 'negative' mould of the user's ear canal. The impression of the ear is made in the office of the hearing health care professional and is out of the manufacturer's control.

Once the physical impression is shipped to the manufacturer, the process of fabricating a hearing instrument can begin. This process is performed by highly skilled workers in a manually intensive series of operations, which result in the impression being transformed into the shape of the desired instrument. The final instrument shell is typically formed from an acrylic material that is cured using ultraviolet light in a clear mould. Afterwards, the electronics module is attached and the instrument is tested and delivered to the customer.

The entire process presents many opportunities for errors to be introduced into the final product. The hearing health care professional may be hesitant to force impression material far down the ear canal of the patient. A deep impression is necessary for a good-fitting, small instrument. However, many patients find it uncomfortable, and removing the deep impression can cause discomfort. The impression can have bubbles or imperfections and can be distorted if it is removed before it is fully cured. Each moulds and impression also has some shrinkage. Because the sculpting of the impression is performed in a custom manner, the skill of the given technician weighs heavily into the final result – therefore one can say that not all shells are created equal.

Overall, due to this high amount of hands-on customization the process is difficult to control and is highly dependent on the abilities of the given technician. Therefore the outcome is not consistent from one instrument to the next.

## 12.3 Digital Manufacturing

Computer aided design and manufacturing (CAD/CAM) of hearing instruments is the current innovative approach to shell manufacturing in the hearing aid industry (see Figure 12.1). This evolution in hearing instrument manufacturing has been the result of unprecedented technological advancement in three-dimensional scanning, automated production hardware and three-dimensional modeling software. Harnessing these advances for the production of hearing instruments that provide optimal customer benefit and satisfaction is now the thrust of the industry.

Over the past years, software applications have permeated aspects of the hearing aid design and manufacturing process, with the notable exception of shell production. Advances in software applications are very noticeable in programmable hearing instruments and especially in the signal processing of the acoustic input/output. Applications less obvious to the dispenser and the consumer include acoustical and electromechanical design and testing, and mechanical design of components of the hearing instruments and peripheries.

Digital Manufacturing technology streamlines this previously manually intensive process and integrates advanced design automation and Rapid Prototyping (RP) technology into a single process. This technology combines

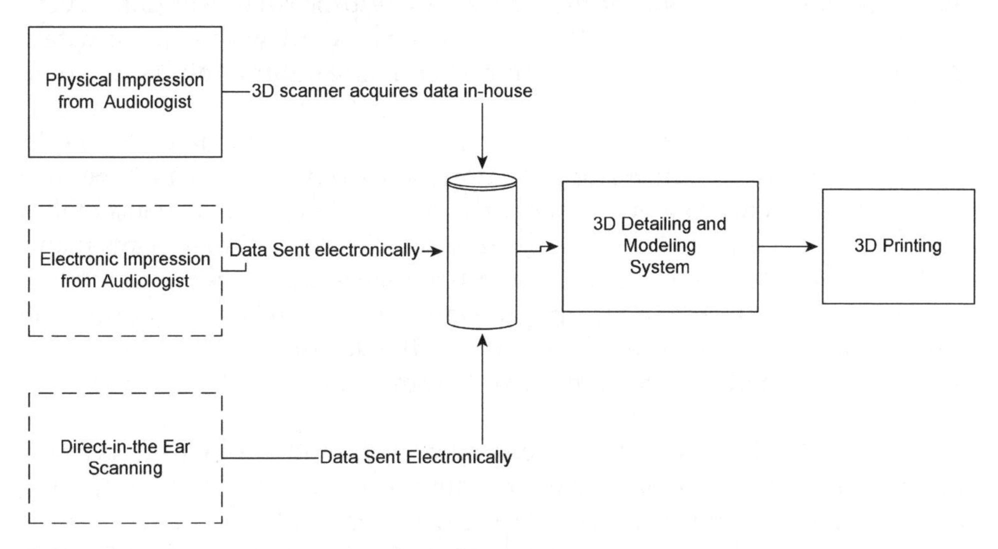


**Figure 12.1** A simplified model of an integrated hearing aid manufacturing protocol. (Copyright Siemens Hearing Instruments Inc., USA)

innovative three-dimensional modeling software development, scanning technology and advances in three-dimensional laser sintering. The result of this combination of technology is the most accurate replication of the ear impression that has ever been available for the purposes of producing a custom-made hearing aid.

The manufacturing process is divided into modeling and manufacturing. The modeling aspect includes CAD modeling (electronic detailing and modeling) of the impression and the manufacturing aspects include three-dimensional printing, finishing, quality control and final assembly.

## 12.4 Scanning

The main prerequisite for the success of the new RP manufacturing process for hearing instruments is a good digital data set to define the geometry of the user's outer ear. The new process does not use moulds; instead, the ear impression is used to create a digital data set through laser scanning. In this approach, the impression is taken as described above and then sent to the manufacturer where it is scanned using a laser digitizer.

While there are a number of digitizers available today, the ear impression is a particularly difficult object to image without missing some areas because of the convoluted nature of the outer ear. The impression must be viewed in 360° azimuth but it must also be viewed in at least two angles in the vertical orientation to view the crevices within the impression.

The sampling rate of the scanner determines the accuracy of the replication of the ear and the size of the data set to be processed. Sampling every 0.1 mm yields a resolution of 0.2 mm. In an average ear with approximately 1200 $mm^2$ of surface to scan, a 0.01 mm scanning sampling results in a data set of 120 000 points.

Scanners in the manufacturing facility must accommodate the rapid processing of large numbers of ear impressions daily. Scanners have also been deployed in hearing instrument dispenser's offices. The dispenser takes an impression of a patient's ear in the traditional way, and then scans it and sends it to the manufacturing site. The advantage to the dispenser is that the order is processed faster and the impression can be stored electronically. The digital data from the local dispenser or the factory scan are similarly modified using software to transform it from a model of the patient's ear to a hearing instrument.

Existing technology, specifically, computerized tomography (CT), is available to capture the ear geometry by scanning from outside the head. In a CT scan, the air–tissue interface seen in the ear canal and pinna areas is well defined and can be readily segmented to reconstruct the surface (see Figure 12.2). In the example shown here, the point cloud is extracted from

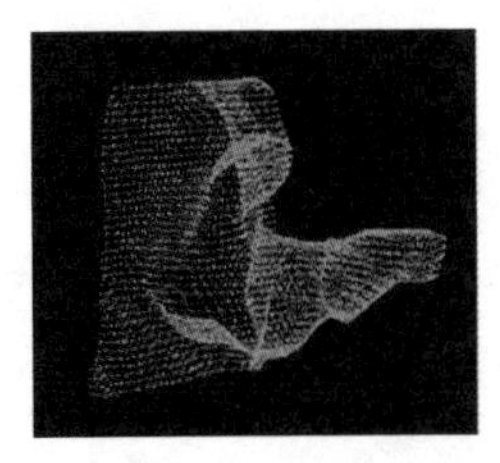
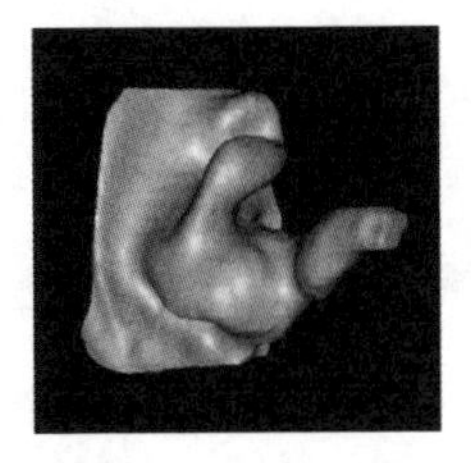
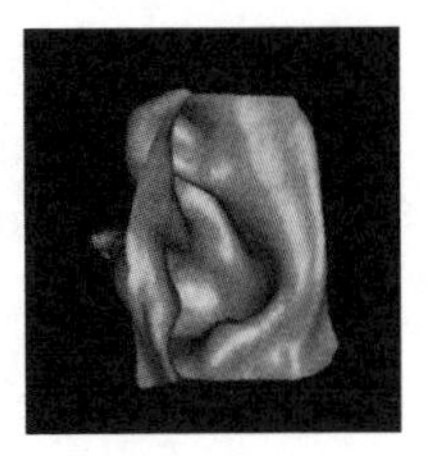

**Figure 12.2** Segmentation of the external ear anatomy from a CT scan. (Copyright Siemens Hearing Instruments Inc., USA)

the CT scan. The middle image is the surface created from the point cloud and the lower image is the same image rotated to view the pinna. To demonstrate the viability of using CT scanning, well-fitting hearing instruments were produced from CT scan data.

## 12.5 Electronic Detailing

Electronic detailing includes the definitive reduction of an ear moulds impression to a device type. Ear impression moulds are scanned using three-dimensional laser scanners. This exact electronic replication of the ear mould is digitally reduced to a device type and designed to dispenser specifications (device model, instrument type and shell style). Figure 12.3 shows an undetailed impression received from the dispenser. The three-dimensional scanned data (point cloud) of the impression is also given in Figure 12.4. The point cloud is subsequently triangulated and modeled into the required device type (STL, or Standard Triangulation Language, file).

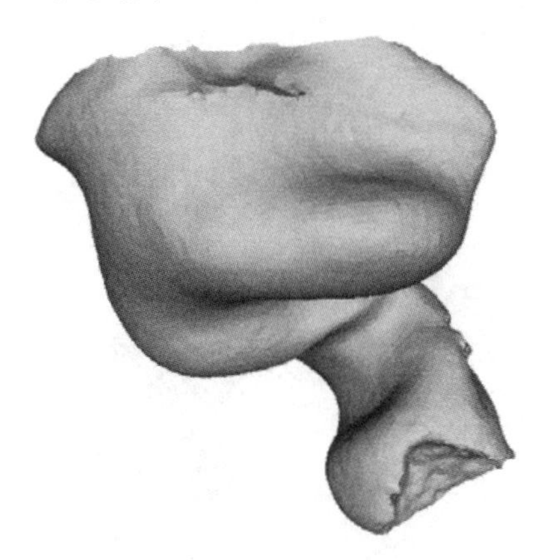

**Figure 12.3** The as-received ear impression from the dispenser. (Copyright Siemens Hearing Instruments Inc., USA)

**Figure 12.4** Scanned impression from a three-dimensional scanner. (Copyright Siemens Hearing Instruments Inc., USA)

## 12.6 Electronic Modeling

Detailed shells are then modeled in three-dimensional modeling software packages. Information from the original, reduced impression and the transformation information from the original impression to the reduced impression are electronically preserved. Thus, lost or damaged hearing aids can readily be replicated to the original specifications. Modifications or change requests in device models or instrumentation, if required, do not involve a repeat of the modeling process but a retrieval of an archived electronic version and subsequent replication.

Figures 12.5 and 12.6 show the original impression and the reduced (detailed) impression. Note that in Figure 12.6 the original impression is

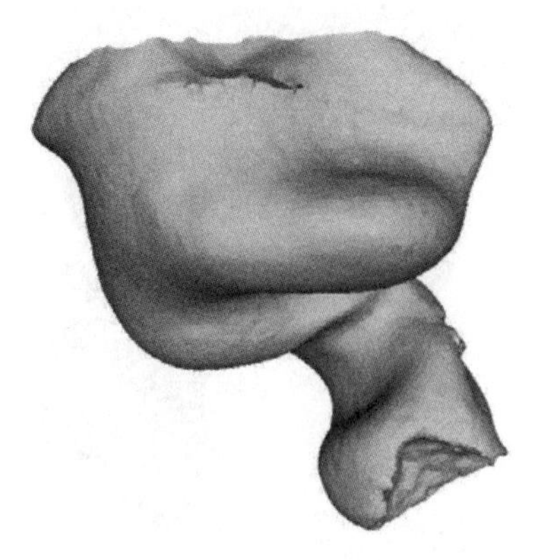

**Figure 12.5** As-received impression from a dispenser. (Copyright Siemens Hearing Instruments Inc., USA)

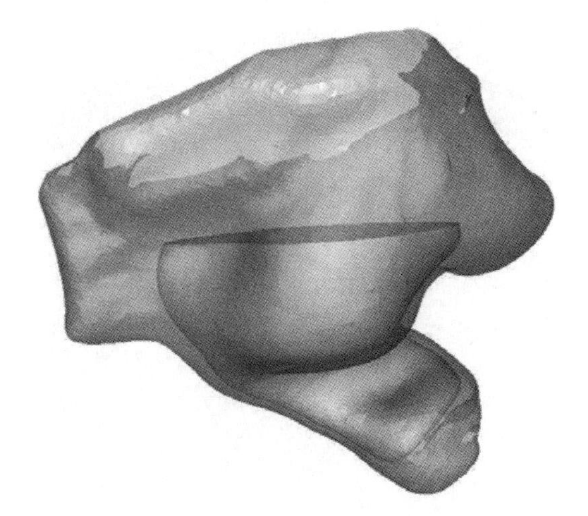

**Figure 12.6** The detailed impression overlaid with the as-received impression. (Copyright Siemens Hearing Instruments Inc., USA)

overlaid on the detailed impression for quality inspection. Subsequently, the device attributes such as vents, receiver holes and other geometric modifications (e.g. bore canal, wax guard, etc.) are added (see Figures 12.7 and 12.8). The completed shell, which is in STL format, is then exported to the three-dimensional printing station.

It is important to note that in this approach the modeling protocols involve the pre-assembly of electronic components in the virtual shell to assess optimal placement and to detect *a priori* which components will fit in the physical world. Thus, instrumentation can be modified electronically prior to manufacturing.

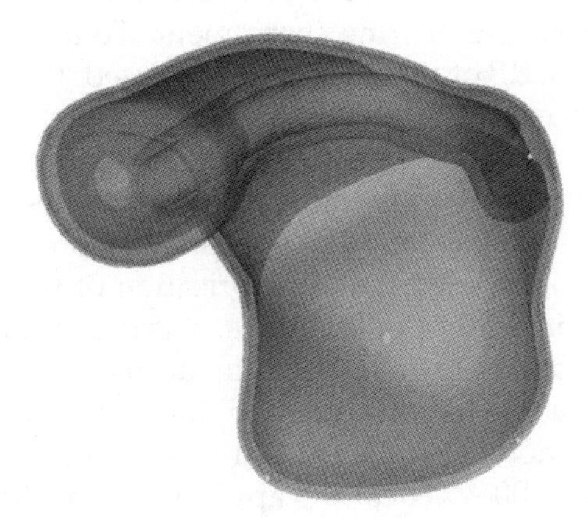

**Figure 12.7** A virtual shell with visualization internally. Geometric collision can be highlighted in this mode. (Copyright Siemens Hearing Instruments Inc., USA)

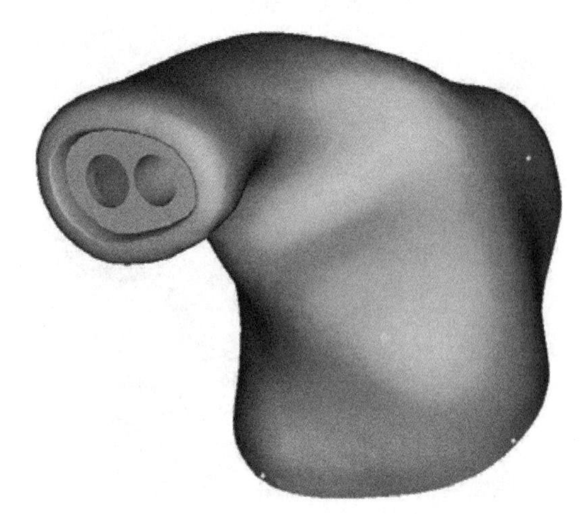

**Figure 12.8** A semi-completed shell with vents and receiver holes included. (Copyright Siemens Hearing Instruments Inc., USA)

Ongoing development activities are directed at harnessing advances in the areas of feature recognition, geometric reasoning and machine vision to automate shell manufacturing protocols. In the areas of feature recognition and geometric reasoning focus is being placed on the development and application of algorithms and definitive methods that will facilitate identification and classification of device types, thus enhancing automated device manufacturing.

## 12.7 Fabrication

Once the scanning and modeling are completed, a shell must be fabricated in order to create a usable hearing instrument. To accomplish this, techniques developed for the RP of moulded parts are used. RP equipment has been available for many years. Prototypes are built directly from CAD models using RP equipment and are typically used in visualization by designers and for marketing evaluations. Until recently, these techniques were not capable of building functional models in any but the least demanding applications, but materials are now available that allow fabrication of parts from a variety of functional materials including polyamides, acrylics and even metals.

There are several attributes of objects made using these techniques that must be carefully considered. Most of these techniques use additive processes that build models one layer at a time. Layer thickness is typically 100 to 150 μm (0.004–0.006 inches) for applications in the hearing industry. This layering effect is readily apparent on areas of a model that have large flat areas parallel to the building plane, so post-processing techniques must be employed to create a smooth surface if one is desired. This is problematic if a very accurate model is desired. The need for a good acoustic seal makes that extremely important in the hearing industry.

The accuracy and feature size that is achievable using RP technologies is also an important consideration. While the accuracy and repeatability of RP equipment has improved, it does not compare with the performance of moulded parts. RP techniques can create features that are on the order of 75 μm in some cases but most are 250 μm or more, and are able to hold tolerances of ± 150 μm. Moulded parts can have features that are 25–50 μm in size with tolerances in the ± 5–10 μm range. This means that in many cases, moulded parts are not compatible with RP technologies.

## 12.8 Equipment

There are a variety of RP techniques used today, but not all are suitable for hearing aid fabrication. Most of the reasons for their unsuitability are material related. The environment of the human ear is harsh. Sweat, humidity, ultraviolet (UV) exposure and cerumen (earwax) can all have strong effects on materials. The effect of materials on humans is also a concern. There are two approaches that are currently being used to fabricate hearing aids in the world market today. The first uses selective laser sintering (SLS) equipment; the second uses stereolithography apparatus (SLA). Both of these systems are available from 3D Systems, Valencia, California, USA. The SLS process was developed by DTM Corporation in Austin, Texas in cooperation with the University of Texas, and the SL process was developed by 3D Systems. SLS equipment and materials are also available through EOS Systems of Munich, Germany. A new process, which has been called raster-based processing, developed by EnvisionTec GmbH of Marl, Germany, is also promising.

## 12.9 Selective Laser Sintering (SLS)

SLS is a process that creates physical models by using heat to melt powdered materials in precisely controlled locations. The materials that work best in the process are those with semi-crystalline structures that have a distinct melt point. The process works by heating a chamber full of powder to just below its melt point and then using a laser to give very localized areas the remaining energy necessary to melt. This allows the powder granules to fuse with the surrounding granules and also the previously fused layers underneath. The next layer of powder is then deposited on top of the built layers. Typically, the built layers are lowered stepwise into a chamber so that the fresh powder is deposited and smoothed into a flat surface.

The biggest advantage to the SLS process as it is used in the hearing industry is the material. When shells were first made for the hearing industry using RP technology, the only biologically compatible material available was the polyamide Duraform PA ® material available from DTM Corporation in Austin, Texas. It was to be years before stable, biocompatible resins for SLA were available. The material properties of polyamide powders are discussed more fully in section 12.12 on materials.

Another advantage to the SLS process is the resulting surface texture. The SLS process can be adjusted until the surface texture has a matte finish, which is very effective in hearing instruments. One reason for this is that the granular nature of the materials does not tend to create sharp edges. Clinical trials found that one of the main advantages of the SLS process hearing aid shells over the traditional UV process was retention in the patient's ear. UV shells have always been made to be very smooth and shiny, and while this is very attractive, they can slide around in the user's ear. The shape of the ear canal changes when the jaw moves during talking or chewing. This can dislodge the hearing instrument and allow it to produce feedback that results in an undesirable loud squeal. The textured surface of the SLS shells reduces the occurrence of this problem.

The main disadvantage of the SLS process, from a design perspective, is that it is not able to make small, precise features. This is a problem if one would like to create features on a hearing aid shell that could be used to snap in moulded components. This situation stems from a combination of the granular nature of the powder and the precision of the laser pointing systems. Walls and features less than 0.5 mm cannot be made with SLS technology as it exists today.

Economically, the equipment needed to operate an SLS system is large and expensive. Smaller systems that are less expensive and have similar throughputs are becoming available to compete with the SLS systems. SLS systems have another advantage here though, because they can be loaded with several layers of shells and allowed to run over the weekend or overnight and will run unattended. Some of the smaller systems have a 2–3 hour cycle time, but an operator must be available to load and unload parts.

## 12.10 Stereolithography Apparatus (SLA)

Stereolithography (SL) is a process that creates physical models through the interaction of photosensitive resins and lasers. Resins have been developed that solidify when they are exposed to certain frequencies of light. Typically these resins are tailored to react to light with wavelengths in the range of 300–400 nm. A vat of resin is resident in the SLA which also has a UV laser.

As the laser is directed in a path on the surface of the resin, the resin is solidified wherever the beam touches. In this manner, through a series of interwoven laser paths, a layer of the object desired, in this case a hearing aid shell, is formed.

There are several issues that need to be handled correctly in the case of manufacturing hearing aids using the SLA. One important issue is the generation of supports. Since the fabrication process is done one layer at a time, some means of anchoring the devices is required. These anchors are called supports and resemble scaffolds used in building construction. Identifying the location of the lowest point on the shell is critical. Unless a support is generated at that precise spot, the first layer will float away. This will result in either an error in the shell geometry or the failure of the process to create a shell. The failed shell can then migrate in the vat, causing other shells to fail or have unwanted layers incorporated into them, which are both catastrophic to the desired outcome. In extreme cases, the entire batch of shells can 'crash'. Finding the lowest point on each shell is difficult when hearing aids are being built because the geometry of each ear is so unique. The geometry of the lowest point can vary widely from person to person.

Another issue is build orientation. Due to the layering inherent in the process, a smooth curve is only approximated with any rapid prototyping process. In areas that have gently rounded curves that are parallel to the build plane, this results in relatively large flat spots that could be described as terraces or contour intervals as seen in topographical maps. These surfaces can be divided for discussion purposes into two categories. Surfaces are either facing up out of the vat of resin or down into the vat of resin. The surfaces of the two layers are markedly different in appearance. The surface that faces up is much smoother and flatter than the down-facing surface. This is readily apparent under inspection since the up-facing surface is reflective and shiny. The shiny surfaces make the layers more obvious and are therefore cosmetically undesirable. The edges of the up-facing surface are also sharper and more precise. This is an advantage if crisp, precise features are desired, but a disadvantage if the surface is desired to be smooth.

Surface texture is also a problem. Traditionally, hearing instruments have been smooth and shiny, presenting a very attractive, lacquered appearance. While this is not necessarily a clinical benefit, it has a desirable cosmetic appearance. Clinical trials have actually shown that a more textured surface is better for retention in the user's ear canal, but hearing aids are expensive and the cosmetic appearance is very important. This leads to the requirement of some post-processing operation. The shells must either be coated with a lacquer of some kind or polished to make them smooth.

## 12.11 Raster-Based Manufacturing

Recently, a new type of Rapid Manufacturing (RM) equipment has become available which has the potential for use in the fabrication of hearing aid shells. This equipment was developed by EnvisionTec GmbH of Marl, Germany, which is now a subsidiary of EnvisionTec Inc. of Detroit, Michigan, USA. This equipment uses photoinitiated acrylic resins very similar to the SLA resins. The difference is that the EnvisionTec equipment uses a projector similar to the ones used to project presentations and slide shows. The image is directed on to a plate with a thin layer of resin through a lens. The photoreactive resin is solidified by exposure to the light and then the process repeats with another layer being applied.

The advantages of this system are its simplicity and speed. The only moving parts are related to the movement of the stage. In an SLA or SLS system there are multiple actuators that direct the laser, roll, wipe, raise and lower parts of the system. All of the moving parts have to be calibrated and maintained, or the parts will not be built accurately. With this new system, the only motion is up and down. In addition, the SLA and SLS systems require a laser to draw the layers of the parts as described above. Parts with thick sections take a great deal of time to build in these systems. The EnvisionTec approach creates an entire layer each time a slide is projected. The exposure time is a few seconds per layer. This can result in increased build speeds on the order of an inch per hour, which is much faster than the other systems. The thickness of the layers is much less than the SLS or SLA processes are capable of. Layers can actually be made down to 20 μm with some resins.

The drawbacks to this approach are that the build area currently available is very small if a high resolution is desired. The projectors have a resolution of around 1200 by 1000 pixels, so if a resolution of 100 μm is needed, the build area is only 12 cm by 10 cm. Newer systems are able to deliver higher accuracy with larger build areas, but they are not yet available in production quantities. The nature of the process is that the three-dimensional object will also have a pixilated nature. This means that along with the layered effect, objects will have a 'Lego Block' appearance. This must be considered if fine details are required. The orientation of the object will change the look of the pixelation. For example, a flat surface that is parallel to the axes of the pixels will be flat, while one at an angle will have a stair step appearance.

The EnvisionTec process requires that the object be built in contact with a clear plate. When the plate is moved, the part must be separated from this plate, so that a fresh layer of material can be introduced. This removal force is significant and requires a large number of very strong supports to be built on the desired part, in this case a hearing aid shell. These supports are very similar to the SLA supports described above. Removal of the supports can leave remnants on the shell that must be removed.

## 12.12 Materials

The first materials available for use as hearing aid shells were the polyamide SLS materials. At the time, there were no other biocompatible resins available that were capable of making hearing instrument shells. Polyamide is a tough, durable resin widely used in consumer products. It was originally introduced under the well-known trade name Nylon®. The SLS process takes powdered polyamide and fuses it into larger objects. The material never receives enough energy to form free radicals or break the polymer chains. The result of this is excellent biocompatibility. The polyamide hearing aid shells are very tough and impact resistant, and testing has shown that they survive the harsh environment of the ear very well. The environmental resistance of the polyamide powders presents another difficulty, however, because they are difficult to bond. The manufacturing process uses special adhesives that were chosen for this application after extensive research.

Hearing instruments fabricated using the SLS process and polyamide powders have been manufactured for several years with hundreds of thousands in the field today. Their use has been well received, and their success has proven that the RM techniques used to fabricate them are clinically and economically viable. The surface texture inherent to the SLS process has been shown to be clinically advantageous because retention in the ear is better than the traditional manufacturing process. The toughness of the polyamide shells has been proven, because, while it is common to receive hearing aid shells for repair that have been cracked or shattered when stepped or sat upon, no polyamide shells have ever shattered to the knowledge of the authors.

The main disadvantage to the SLS process is, ironically, due to its toughness. The dispensers of hearing aids have traditionally modified the instruments if necessary to fit better in their patients' ears. An example of this would be grinding away some of the shell where a patient feels it is too tight. The polyamide shells can be modified, but it is more difficult than with an acrylic, so new techniques and materials have to be used by the dispenser. While this is a manageable situation, it has also been used as an argument for the use of the emerging SLA hearing aid manufacturing process.

The first materials that were developed for SL in RP were acrylic resins. More recently, epoxies are being used for their higher strength and dimensional stability. Unfortunately, the epoxies are not biocompatible, so while they may be good for making marketing samples of consumer products like telephones, they are not suitable for prolonged contact with human tissue. The acrylics were found to distort large models with thick sections when they cure, which is why the epoxies were developed. This is not a problem for hearing aid shells, since the models are small and the walls are thin. The acrylics used are very similar in chemistry and behavior to the materials that the hearing aid industry has used for decades. Originally, the hearing aid industry used a heat-curing process, but within the last 10 years UV curable acrylics have been used that are similar enough to the RP resins that the RP resins will actually work in the manual hearing aid process. The similarity of materials is a big advantage for the SLA process since hearing care professionals commonly modify hearing aids in their offices so that they fit better in the customer's ear. The techniques, materials and equipment used to modify SLA shells are almost identical to traditional methods.

SLA resins are also similar to the UV curable resins in the fact that they are very brittle. This is not an issue in normal day-to-day use, but it causes problems when manufacturing and repairing hearing aids. SLS materials, on the other hand, are much more robust and they do not shatter.

The materials used by the EnvisionTec equipment are acrylic-based resins that are in the same family as the SLA resins and the traditional hearing industry manual processes. This gives them many of the same advantages and disadvantages as possessed by those materials. The EnvisionTec process is different in one key aspect because it uses light in the visible spectrum. This means that there is much less energy available for use in curing resins. Among other things that this can cause, it can make control of the layer thickness difficult. The thickness of layers is largely set by the motion of the build platform, but when an area such as a small overhang is desired, the thickness is determined by the depth of the penetration of the radiant energy. This results in less precisely formed features on the back side of parts made using this process.

## 12.13 Conclusion

The hearing instrument industry has started evolving from a completely manual process to a digital process. Digital Manufacturing is particularly well suited to the hearing industry because of the need to fit each user's ear with a custom appliance for maximum performance. The change has been in progress for the past four years, starting with the introduction of hearing instruments that were fabricated from polyamide powder using the process of SLS. However, new materials and techniques are being developed constantly, so that today there are many more possibilities for use in manufacturing hearing instruments. The evolution has required development of new software and drastic changes in the paradigms that rule the hearing industry. However, it offers new opportunities to control the quality and cost of the traditionally manual process.

More drastic changes will come to the hearing industry when the trends set in motion come to fruition. Today it is already possible for the dispenser to scan the impression of a patient's ear in their office and send the information electronically to the manufacturer. However, the physical impression is still required. This is still a source of error, takes time from the dispenser and is potentially injurious to the patient if a deep impression is required. In the future it will be possible to directly scan a patient's ear and send the information electronically to the manufacturer. When that day comes, the manufacturing process will no longer have the physical impression to work from and the only way to build an instrument will be by using Digital Manufacturing techniques.

# 13

# Automotive Applications

Graham Tromans
*Loughborough University*

## 13.1 Introduction

Of all the potential areas of application for Rapid Manufacturing (RM), the automotive industry probably offers the most significant opportunities for changes in the way manufacturing is carried out. The design constraints currently imposed on the automotive designer owing to tooling design limitations will be removed. In both high and low-volume niche market models, the ability to individually customise areas of the car to suit customers' requirements would have a particular impact in areas such as ergonomics, where parts could be manufactured to make the overall comfort fit of the car suit the customers' needs. At the moment the ability to do this is very limited due to restrictions imposed by tooling. With the linking of technologies such as scanning and additive manufacturing, customisation becomes very real.

This chapter aims to highlight some of the applications already used in the automotive industry, and will cover areas primarily dealing with motor sports. The cross-section of applications highlighted here is just the start of what promises to be a complete new set of manufacturing processes for the automotive industry. The high added value and low-volume nature of motor sports provides a good fit with the current technologies that are available.

As RM processes become quicker and cheaper it is likely that the knowledge and understanding gained in motor sports will filter through to other areas of the automotive industry. Indeed, the concept of applying initial development in the motor sports industry with a view to filtering the technology through to the production car industry is well established. A

*Rapid Manufacturing: An Industrial Revolution for the Digital Age*
Editors N. Hopkinson, R.J.M. Hague and P.M. Dickens 

topical example of this may be found in a UK Government funded study that reports on the feasibility of using developments in the motor sports industry to improve energy efficiency in the automotive industry as a whole [1]. The view behind the feasibility study is that many of the needs in motor sports, such as efficient fuel combustion, are also required 'down the automotive chain' in production cars and that a transfer of technology from motor sports to production cars will prove to be an efficient use of research resources.

Infact, within RM, this filtering 'down the automotive chain' is already happening, with many production car manufacturers currently making use of RM technology for end-use products, albeit in specialised cases. Many perceive RM technologies as holding a key competitive advantage; for this reason many of the automotive manufacturing companies are very guarded in what information they release. This is highlighted in the lack of production vehicle case studies available for publication in this chapter.

## 13.2 Formula 1

The use of the Rapid Prototyping (RP) technologies has, for some time now, been commonplace in the development of Formula 1 (F1) racing cars. With the advent of materials with better functional properties it did not take long for this industry to realise that these parts could be fitted directly to the cars, thus allowing quicker modifications to be made at a greatly reduced cost.

One of the pioneers of this type of application was the Renault F1 team who decided to establish a partnership with 3D Systems Inc. and set up an Advanced Digital Manufacturing Centre, at Enstone, in the heartland of the British motor sports industry in the UK. This facility was created to support the team requirements as far as design and supplying parts that were to be made on the RP machines and fitted directly to the car (RM). Figure 13.1

**Figure 13.1** The Renault F1 car, which includes a number of parts made by Rapid Manufacturing. (Courtesy of Renault F1 Team)

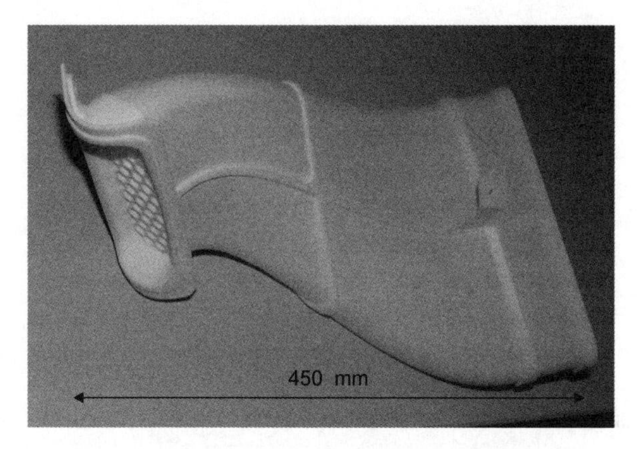


**Figure 13.2** Cooling duct. (Courtesy of Renault F1 Team)

shows the Renault F1 car, which includes a number of parts made by RM.

A major benefit of the ability to manufacture parts directly from these machines was the lack of manufacturing design constraints. This allowed assemblies of parts to be reduced into single components.

## 13.3 Cooling Duct

This duct (see Figure 13.2) was designed to improve the cooling of the electrical system. The nature of the geometry makes it impossible to manufacture in one piece using traditional manufacturing techniques such as carbon fibre lay-up. This complex geometry is necessary because of the packaging problems of a modern F1 car. The duct, which is made from selective laser sintered nylon-12, is an on-car component and is used in every race. The duct was subject to a number of design changes as the car evolved, making full use of the RM technologies available.

## 13.4 The 'Flickscab'

This part (see Figure 13.3) was an aerodynamic improvement found in the wind tunnel just prior to the French Grand Prix, which was to take place that very same week. It was found that by thickening the section of the front wing end-plate flick there was a substantial gain. It was decided that to remanufacture the front wing end-plate assembly would be impossible in the time, so a shape-changing scab would be used. The part was designed at Renault's facility on the Thursday morning prior to the race, built on one of

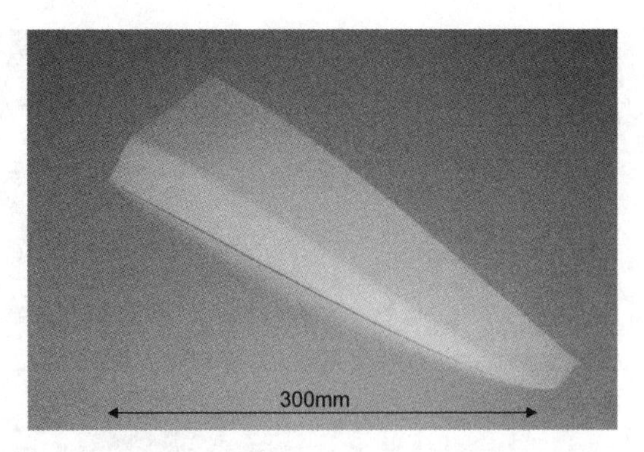


**Figure 13.3** The 'Flickscab'. (Courtesy of Renault F1 Team)

the selective laser sintering (SLS) machines on Friday using nylon powder, followed by removal from the machine and post-processing at 3.00 am on the Saturday. It was then hand-delivered to London where it was put on a plane for Paris. The part was fitted to the car on the Saturday morning in time for qualifying, and then raced on the Sunday.

The examples above from Renault highlight two of the main benefits that may be obtained from Rapid Manufacture, namely the ability to produce complex geometries (cooling duct) and the ability to quickly turn around new or replacement parts (flickscab). Although one may say that these applications are more like RP, it only goes to prove that parts can be manufactured on these types of systems and be used in a highly demanding real-world environment.

Other F1 teams have openly disclosed the fact that they use parts made on RP machines on their race cars. Indeed, as far back as 2001 it was reported that Jordan were using over 20 such parts on each race car. Although Jordan owned two stereolithography apparatus (SLA) machines themselves at that time, it became obvious to them that by using the SLS process, they could fit parts directly on to the car and use them under race conditions. In conjunction with a service provider, the three-dimensional Rapid Prototyping/ Tooling (3DRPT) based in Berkshire, UK, they produced a connection box that was traditionally manufactured by hand-laying carbon fibre into tooling, a very time-consuming and laborious process, taking a number of days, which includes the manufacture of the tooling. The use of SLS, however, meant that it was possible to manufacture around 12 connection boxes in the time required to make one by hand-laying carbon fibre. Encouraged by the success of this technique the engineers at Jordan looked at other areas of the car where the technologies could be applied. This resulted in areas such as aerodynamic parts and cooling ducts being produced and fitted directly to the race car. More recently body panels of around 200 mm $\times$ 300 mm have been made and fitted [2].

Toyota Motorsport also use RM type processes, including SLS from EOS [3], to form parts that are fitted directly to the race cars. However, as in other race teams, due to the competitive nature of F1, little detailed information in terms of applications is publicly available.

## 13.5 NASCAR

Another area of motor racing that saw the potential benefits of using this technology to produce functional parts that could be fitted directly to the race cars was NASCAR (National Association for Stock Car Auto Racing). The use of the technologies has helped in producing many parts that were used during a race, such as airflow parts and brake air inlet ducts. As with the Renault F1 team, Penske, who are based in the USA, had the foresight to set up an Advanced Digital Manufacturing Centre in conjunction with 3D Systems Inc. Many parts have been manufactured in the RP/RM facilities owned and run by NASCAR teams throughout the USA, which have been fitted directly to the cars and used during races.

## 13.6 Formula Student

Formula Student is an annual international automotive racing competition that has been running since 2001 and includes teams from universities worldwide. Because the Formula Student project runs during the academic year, the design-to-track time is very limited and so to achieve the ambitious design objectives the team make extensive use of the RM facilities available at the university. The key aspect of the technology as far as the team is concerned is that it allows more design time, which gives further opportunities for refinement, which, because of the flexibility of the processes, can be done without compromise.

Initially the team used a 3D Systems SLA7000 machine to produce some relatively simple components – primarily as a straight swap for traditional fabrication techniques. As the potential of the technology became clear, the design of successive cars has incorporated parts designed specifically for RM.

For the 2004 season, the car was designed with tight packaging constraints in order to reduce both overall mass and to keep the centre of mass as close to the ground as possible. The bodywork was a development of the previous car and featured many Rapid Manufactured components for sections where intricate details would make conventional moulding difficult. The overall shape for the body was designed using the Solid Edge computer aided

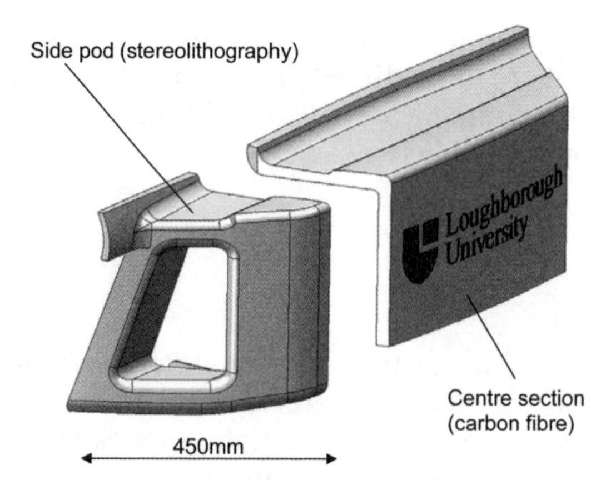


**Figure 13.4** The left-hand side pod of the Loughborough University Formula Student car. (Courtesy of Loughborough University)

design (CAD) program and the complete shell then split into various sections whereby some would be manufactured using traditional moulding techniques such as carbon fibre lay-up and some, such as the side pod end details, were manufactured using the SLA7000 machine.

Figure 13.4 shows the left-hand side pod where the front end detail (SLA component) is bonded to the centre section (carbon fibre moulding). The end detail is constructed as a single model and includes a return flange to allow for easy bonding to the centre section.

A similar approach has been taken with the cockpit section whereby the air scoop and head fairing is an SLA component (see Figure 13.5) bonded to the cockpit surround moulding. Also shown is the actual SLA component before being bonded to the surround.

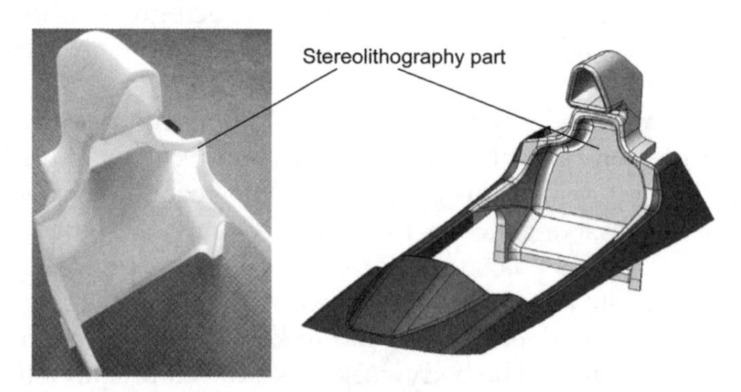


**Figure 13.5** Stereolithography air scoop and head fairing. (Courtesy of Loughborough University)

**Figure 13.6** Exploded view of the bodywork assembly with Rapid Manufactured parts in white. (Courtesy of Loughborough University)

**Figure 13.7** Complete vehicle with Rapid Manufactured parts in white. (Courtesy of Loughborough University)

Figure 13.6 shows an exploded view of the bodywork assembly and Figure 13.7 shows the complete vehicle with the Rapid Manufactured bodywork components shown in white. A total of 28 body parts were manufactured using stereolithography (SL). SL was selected as the main process due to the size of the parts; had they been made by SLS they would have needed to be made in sections and bonded.

The desire to keep the weight as low in the car as possible left little space within the engine bay for the air intake manifold. As part of the engine development program, the inlet pipes of the intake were required to be longer than previous installations. By designing the manifold with RM in

Figure 13.8 Location of the air intake manifold. (Courtesy of Loughborough University)

mind, it was possible to achieve the packaging and performance objectives of the intake. Because of the higher temperatures around the engine, it was decided to use a Vanguard SLS machine to produce a more durable nylon component. The complexity of the model would have been extremely difficult and time consuming to realise with any other manufacturing technique. Using SLS, the part was designed, built and fitted within the space of four days. Figure 13.8 shows the inlet system installed in the car (circled).

Figure 13.9 shows the CAD model, highlighting the complexity of the geometry and Figure 13.10 shows the final component as it is assembled to the car. The complete car (see Figure 13.11) included 28 different parts made by stereolithography, 7 parts made by selective laser sintering and 1 part made using fused deposition modelling (FDM), a total of 36 parts altogether.

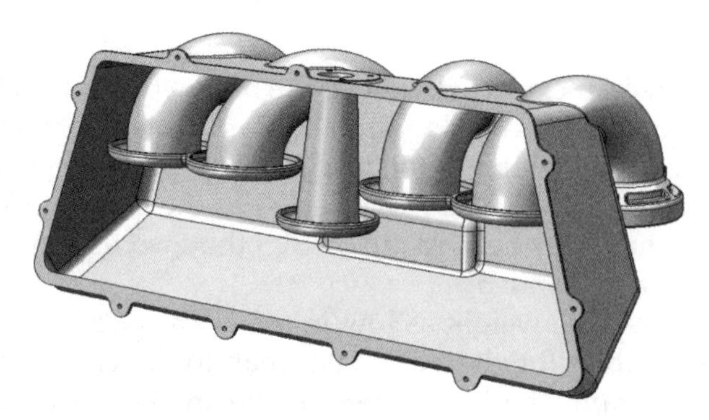

Figure 13.9 CAD model of the air intake manifold. (Courtesy of Loughborough University)

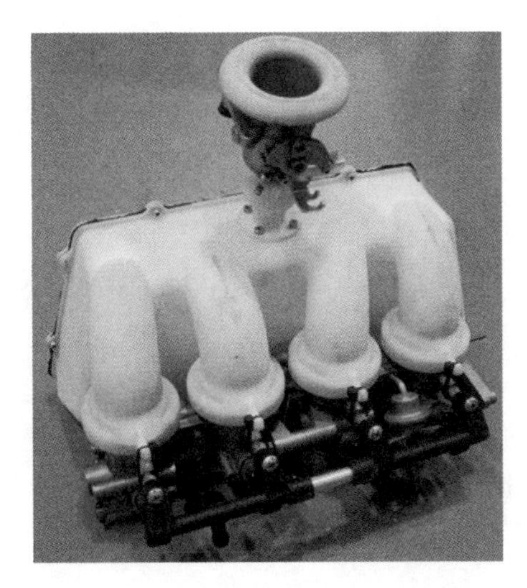

**Figure 13.10** The air intake as it is manifold-fitted to the car. (Courtesy of Loughborough University)

**Figure 13.11** The Loughborough University Formula Student car 2004. (Courtesy of Loughborough University)

## References


1. Energy Efficient Motorsport (2002) A feasibility study into the role motorsport can play in the development of energy efficient automotive technologies, Motor Industry Association.
2. Plunket, T. Racing ahead of the competition (2001) *Time Compression Technologies Magazine*, **9**(4) 25, August 2001.
3. Stocker, M. (2002) From rapid prototyping to rapid manufacturing, *AutoTechnology*, **2**, 38–40.

# 14

# Rapid Manufacture in the Aeronautical Industry

Brad Fox
*The Center for Rapid Manufacturing*

## 14.1 Opportunity

Rapid Manufacturing (RM) has begun to make its way into the aeronautical industry and is set to have profound implications. Although there are potential huge cost savings, its biggest contribution may be the unleashing of designers to create completely new and innovative systems, applications and vehicles. RM will certainly transform the status quo and deliver exciting results.

This chapter will seek to explain why these opportunities exist by highlighting both the process of how different aeronautical companies implemented RM as well as the benefits they gained.

## 14.2 Overview

The objective of this chapter is to highlight the current state of how RM is being utilized within the aeronautical industry today. The chapter will begin by explaining how Rapid Prototyping (RP) introduced companies to the fabrication of complex parts by using a layered-based construction. Next is an important discussion on the main system drivers of a technology that would meet the aeronautical industry's requirements. Third, is a section on how companies have approached the very difficult task of qualifying the

*Rapid Manufacturing: An Industrial Revolution for the Digital Age*
Editors N. Hopkinson, R.J.M. Hague and P.M. Dickens 

**Figure 14.1** Northrop Grumman GlobalHawk. (Courtesy of Northrop Grumman)

process and product the RM equipment fabricates. In other words, how does a part produced in an RM technology compare to parts produced via a 'traditional' manufacturing process? This last section addresses, we believe, a crucial step to the widespread implementation of RM in the aeronautical industry. The reader will hopefully conclude that at the heart of any successful RM program is the acceptance of the RM technology as a qualified manufacturing process. We will discuss how two companies have approached this problem.

It should also be noted that companies are very early on in the cycle of accepting RM as a manufacturing process. Many real-world case studies are very, very new and, at the time of publication, could not be discussed for security purposes. However, some well-known programs are utilizing RM. Figure 14.1 shows the Northrop Grumman GlobalHawk, just one of many vehicles depending on RM for its development.

## 14.3 Historical Perspective

Similar to most industries, the aeronautical community began a path to RM through the avenue of RP – its core technology. We can refer to it here as 'layer-based fabrication'. In fact, the aeronautical industry was one of the early adopters of RP with Boeing and others purchasing equipment in the early 1990s.

As widespread implementation of solid modelling computer aided design (CAD) software took hold and RP processes matured, designers and engineers were introduced to the concept of 'art to part' in days, not weeks. For aeronautical companies, RP began to have a dramatic impact on compressing design times and program costs, to the point today where RP is viewed as just the 'normal' way of progressing through a design build.

As the technology earned its keep by producing cost-effective prototypes, people began to find innovative means to utilize RP past the relatively mundane applications of show and tell parts and design verification. For example, in 2001 Northrop Grumman, the long-time user of RP, won first place in the North American Stereolithography Users Group competition with a novel Field Service Kit. This Field Service Kit used stereolithography to manufacture a series of 'tools' that guided a technician making highly accurate changes to the complex titanium structure of the aircraft.

Other RP-driven applications in the industry were: drill blankets, models for wind tunnel testing, master patterns for investment casting of metal parts, etc. By 2002, RP had found itself fully entrenched as an everyday 'tool' and was depended upon by almost every contractor and supplier in the industry.

Figure 14.2 depicts how Northrop Grumman institutionalized RP and went on to develop the use of RP technology for RM. This figure also communicates the future direction for the implementation of RM.

With the kind of interest RP was generating at Northrop Grumman, as well as almost all aeronautical companies, the seed for asking a higher-level question was sown: 'Would layered-based technologies one day become a means of production?' In an environment as harsh as the aeronautical are, believers were very far and few between! However, this would not stop the early adopters from working hard towards a 'Factory of the Future', as those at Northrop Grumman termed RM.

## 14.4 Aeronautical Requirements for RM

Reacting to the success of prototyping with layer-based fabrication, the manufacturers of 'RP equipment' continued to innovate; some clearly saw an opportunity to realize the progression of RP to RM. The special needs of the aeronautical industry drove them to address two issues. Not only did they need to develop equipment that could be considered capable of true 'manufacturing' but they also needed to develop and introduce materials that could be accepted by aeronautical applications. As a result, not all technologies using layer-based manufacturing leant themselves effectively to RM. It is important for the reader to understand how some important technology and economic drivers (which came out of the aeronautical industry) can limit the processes and technology used.

## 14.5 Why RM Is Uniquely Suited to the Aeronautical Field

To understand why RM has so much potential in the aeronautical field, it will be important to identify and understand these key 'program drivers',

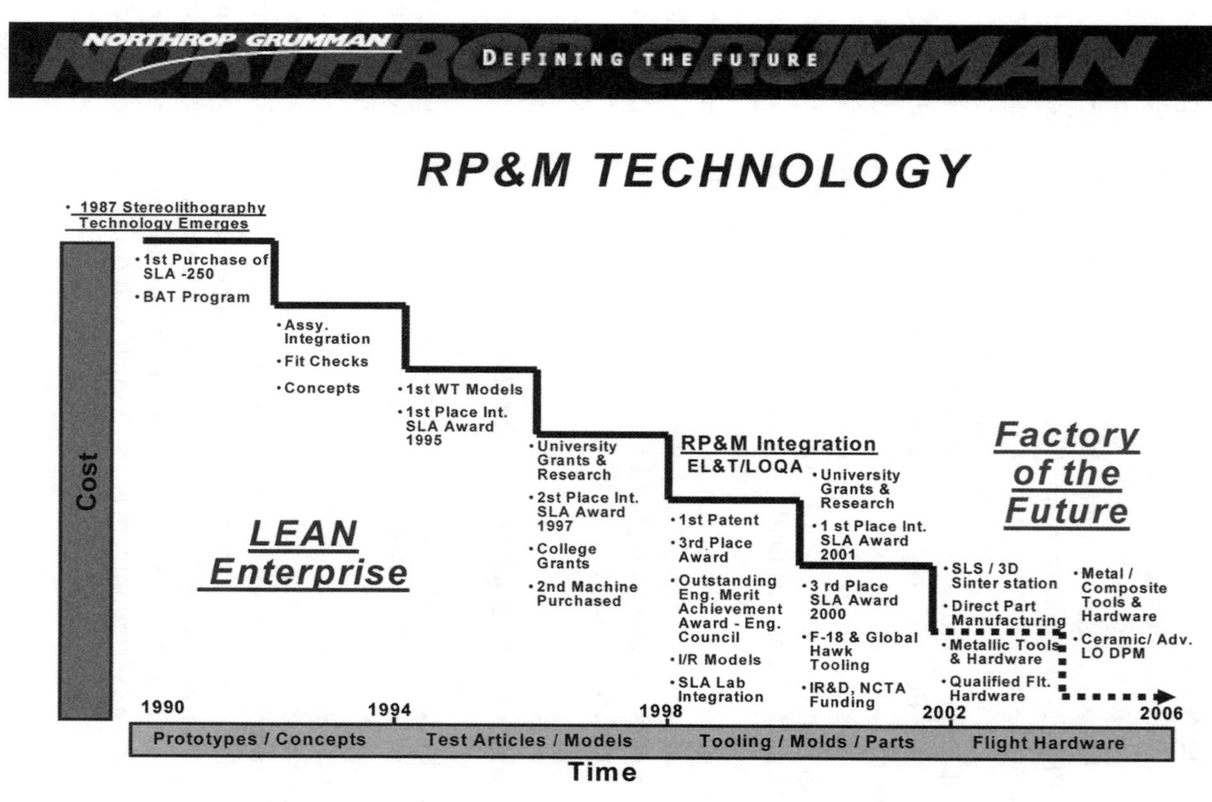


**Figure 14.2** RP and RM technology within Northrop Grumman (1)

i.e. the key measurement parameters that any RM technology must deliver. RM is particularly useful in aerospace for the following reasons:

1. Low production volumes. Aeronautical development has always been costly for several reasons. One of the contributing factors is ineffective amortization of tooling costs over a low production volume.
2. Constant design iterations. Due to such a high number of iterations in design, being able to simply use a new file for the production of the part without any tooling modifications means that RM is capable of huge time and cost savings.
3. Geometric design complexity. The aeronautical industry produces some of the most complex part geometry. As vehicles get smaller and more efficient, the design pressures to combine part numbers together makes manufacturing by traditional means much more of a challenge than when using RM.
4. Testing of new materials. RM has the potential of processing more exotic materials, faster and cheaper. It also allows for prototyping and small-volume trials of new materials, again without the need to ever invest in tooling.

## 14.6 Acceptable Technologies

Not all technologies using layer-based fabrication are a good fit for aeronautical applications. Surveying the (approximately) 10 different layer-based technology platforms that are commercially available today, only two or three are practical candidates for the application in discussion here. They are:

- Selective laser sintering (SLS)
- Stereolithography (SL)
- Fused deposition modeling (FDM)

Chapter 5 discusses details of many different technologies that are being used or may be used for RM and various texts on RP also discuss these technologies at length. However, there should be a short discussion on the strengths/weaknesses of each of the relevant technologies within the context of aeronautical applications. It should also be noted that the aeronautical industry has different requirements for RM and therefore the technology acceptable platforms are limited. Other industries may be able to utilize a wider range of acceptable platforms.

Table 14.1 depicts the three general areas where aeronautical companies are measuring a technology's acceptance level, as applied to RM:

- Materials processed in RM equipment
- Process-related parameters
- Geometry limitations of the RM equipment

Materials should be an obvious and important parameter for RM. Indeed, this has been (and continues to be) one of the most limiting factors for widespread integration of RM. Therefore, equipment manufacturers are teaming up with material developers to introduce new, acceptable materials. For now, there are but a few acceptable materials – primarily nylon – and the industry is building on this foundation. Shortly, the industry should see the availability of flame-retardant nylon, as well as other higher-order, engineering grade materials. It should also be noted, as described in Table 14.1, that just because equipment may use an 'acceptable' material it does not necessarily transmit into an acceptable, production grade part.

A second important parameter is process specifications. A highly practical example is the speed in which a part can be produced, because speed translates directly into cost. Additionally, and again highly practical, is the size of part that can be produced in the equipment. It should be obvious that from a prototyping viewpoint, speed and size of a part (larger parts can be glued together) are not hugely important, but from a production viewpoint it is paramount to success.

The last parameter mentioned in Table 14.1 is referred to as 'geometry related', which primarily means accuracy. Again, from a prototyping standpoint, this may not be critical, but when the application are parts to be used in commercial/defense aircraft, the RM equipment must be compared to the well-established, accurate and repeatable production processes such as plastic injection or rotational molding.

It must be stressed that the statements and conclusions in Table 14.1 relate to the unique application of RM and that all of these technologies are proven and highly used RP solutions. While there certainly could be (and are) other important parameters by which to measure potential RM systems, the three areas mentioned here are the foundations for both technology and methodology acceptance in the aeronautical market place.

Of the three platforms mentioned, SLS seems to be the most widely used technology for aeronautical applications. This is probably because:

1. Integrity of layer-to-layer fusion yields a part most closely resembling other accepted forms of manufacturing, like rotational molding.
2. Very fast throughput (manufacturing productivity) hits the required financial targets; i.e. RM must be competitive to other traditional means of manufacturing.
3. The variety of materials that can be used include the future development of materials (such as flame-retardant nylons and even some more exotic metals important to the aeronautical industry, like titanium).

**Table 14.1** Application profile for Rapid Manufacturing in the aeronautical industry

| | SL | SLS | FDM |
|---|---|---|---|
| Materials: | | | |
| Must be able to use acceptable production grade materials | Current acceptance of SL materials is slow. However, a few companies are in the testing phase for some limited RM applications | SLS proves the most versatile by its ability to process both metals and plastics, in the same unit. Has the ability to achieve parts close to existing standards | FDM strength is the ability to use commercial grade plastics such as acrylonitrate butadiene styrene (ABS), polystyrene (PS), polycarbonate (PC) and polyphenylsulfone (PPSF). However, actual 'processed' material properties must be compared to norms. |
| Possibility of processing 'exotic' materials | While SL cannot process metal directly, the suspension of nanoparticles made from exotic materials are a possibility | Current metals include A6, Inconel and aluminum. Future possibilities include titanium and other exotic metals | Processing exotic materials could be an issue since they have to be in a 'form' (filament) that is accepted by the equipment and can be processed (extruded) successfully |
| Process related: | | | |
| Speed – fast enough to manufacture the parts and be economically feasible | Yes, especially when building multiple parts at the same time. Support removal will add time | Yes, able to build multiple parts at once for high throughput. Ability to 'stack' parts and even add parts while building | No, slowest of all the processes, especially in builds requiring multiple parts at the same time |
| Size – being able to create the part as a whole instead of 'piecing' part together | Good build platform of 20 in × 20 in × 24 in. weld joints with the same material are strong | Good build platform of 11 in × 13 in × 15 in. Larger bed sizes by EOS (27 in × 15 in × 23 in) | Good platform sizes. Piecing parts together does not work very well |
| Geometry related: | | | |
| Accuracy/ repeatability/etc. | Acceptable accuracy, feature definition very good | Acceptable accuracy, feature definition is acceptable | Acceptable accuracy, feature definition not acceptable in some geometry |
| Able to easily create the complex geometries without difficulty | Mostly, but some limitation may occur due to supports | Yes, virtually unlimited, since easy to remove powder supports parts | Somewhat limited due to supports, but water-soluble supports allow more complex geometry |

Some of these reasons will become clearer after the following discussion of how aeronautical companies are qualifying RM systems.

## 14.7 Qualifying RM Systems

Now we will begin what seems to be, without exception, the most difficult task facing the aeronautical companies – the task of qualifying an RM system to be 'of manufacturing competence'. To qualify a system's 'manufacturability' (which is different from the more business related parameters discussed in Table 14.1), the issue is twofold. Firstly, the equipment must be proven to be 'manufacturing' capable. Secondly, the materials being used in the system must be from established manufacturing families (i.e. nylon, ABS, etc.) and parts processed in RM must be equal to parts in the same material, but produced in an established process such as rotational molding.

Figure 14.3 shows a suggested decision tree addressing the issue(s) for implementing RM. Several aeronautical companies are currently at the stage of working on qualifying RM processes so that they can be included in 'specifications' on real world programs. Bluntly put, a buyer must have some kind of assurance that if they order a nylon part processed by RM it will have the same (or highly similar) properties as if it had been manufactured by rotational molding. If this could be established, then the opportunities for RM are greatly multiplied. For this reason, the issue of qualification becomes the most important task.

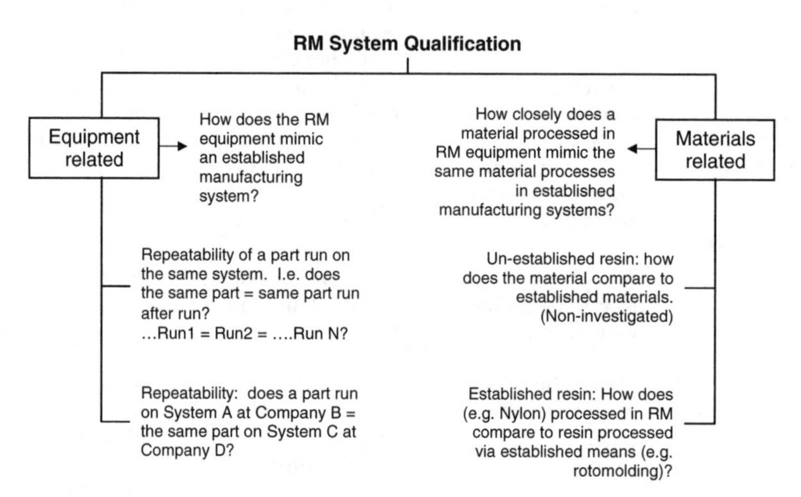


**Figure 14.3** Decision tree for RM in the aeronautical industry

The questions raised in Figure 14.3 are being/have been addressed by all aeronautical companies wanting to implement RM. Some companies, such as British Aerospace, have taken a somewhat 'informal' approach to qualification, while others, such as Northrop Grumman, have taken a more formal approach. These different approaches are discussed below.

### *14.7.1 Qualifying SLS at British Aerospace (BAe)*

At BAe, the opportunity to investigate manufacturing options came when a current program requiring highly complex ducts was targeted for cost reduction. By re-engineering an assembly of over 40 parts down to roughly 26 parts, significant cost avoidance could be achieved. While BAe had experience utilizing SLS for prototyping, they did not know whether the parts could be mass produced using the process – or whether parts fabricated in SLS would be flightworthy. Their quest for answers was to lean on an experienced vendor who had been using SLS for many years.

To qualify the SLS process and equipment, BAe and their vendor decided that physical testing of the parts would be the quickest and least complicated approach. Therefore, a series of parts that formed the ductwork were built on a Vanguard SLS machine. These parts were then assembled and put on to test stands developed by BAe. After a series of testing, re-building of parts with different machine build parameters and implementing post-processing techniques, the parts were eventually certified as flightworthy.

The advantage of this more informal approach to qualification is that it can be done quickly using a very straightforward approach; i.e. do the SLS parts pass the physical testing to qualify them? The disadvantage of this approach is that, more or less, the process has only been qualified for those specific parts, in that specific application. While the testing surely gives some historical data regarding parts built in SLS, any new program would need to go through a similar series of physical testing.

### *14.7.2 Qualifying SLS at Northrop Grumman*

For Northrop Grumman (El Segundo, California) the approach to qualifying SLS was much more empirical and focused on conformity of the system to a manufacturing standard. They felt that a longer-term process of materials testing, and linking the results of the materials testing to machine build parameters, would result in a qualification of the SLS process itself. Only then could Northrop engineers have a level of confidence in this 'new manufacturing process' such that it could be specified broadly across many of their programs.

Deciding on what material parameters to use for qualifying the 'system' was researched at length. Surprisingly, a straightforward answer emerged.

Northrop Grumman's engineers asked a simple question, 'How does a nylon part produced in, say, rotomolding or injection molding compare to a nylon part produced in SLS?' With the base material (nylon) being the same for SLS, injection molding or rotational molding, the difference in mechanical properties between an SLS part and a molded part must therefore occur in the ***processing*** of the material. We know that almost all mechanical properties are greatly affected by the density of the material and we also know that the density of a fused part in SLS has historically always been less than a 'molded' part. Therefore, Northrop Grumman's engineers felt that the primary controlling parameter for qualification became density.

To establish a baseline for density, SLS powder was melted under controlled conditions in a laboratory. Once a solid, the material was machined into a cube of known dimension. After carefully measuring and weighing the cube, the baseline density was now known. Furthermore, this cube became the 'target' density for SLS machines to match. In other words, could an SLS processed cube have equal (or very close to it) density to that of a cube from melted powder under controlled parameters, simulating that of injection or rotational molding?

With this as the rationale, the teams set off to find a set of SLS build parameters that would produce a sintered cube as close to 100 % density (i.e. the control cube) as possible. To find this set of build parameters an initial set of cubes were fabricated in the SLS machine, measured, weighed and the density calculated. Next, an entire array of build parameters was used, tested and measured, in order to understand what parameters effected the ultimate goal of density. Finally, from all of these data, an 'ideal' set of build parameters for the SLS machine was agreed upon.

While this may seem highly straightforward, adding to the complexity was the requirement of having a consistent density across the entire volume of the build chamber (effectively, 11 in × 13 in × 15 in for the SLS Vanguard). Cubes were therefore built throughout the chamber and analyzed.

For those participating in the qualification process, this proved to be no easy task. It was left to each organization to prove a consistent density 'map' across the entire build volume. This meant that, in some cases, the SLS equipment had to be modified to achieve these results. However, without the drive toward this level of conformity, the SLS equipment could not be looked upon as an 'equal' to other established forms of mass production. In the end, the Northrop Grumman engineers felt that, if they could develop equipment and build parameters that could constantly produce density cubes of 99.5 % or better across the build volume, they indeed had qualified the SLS for RM.

As of the writing of this chapter, one organization had been successful in meeting these stringent requirements with the hopes of a second organization to follow shortly.

## 14.8 Summary

This discussion on equipment qualification is meant to highlight a required step to mainstreaming 'layer-based fabrication' for manufacturing. It is not meant to be conclusive or exhaustive, but does point out that each technology platform must be tested to gain the status of 'Rapid Manufacturing'. Other companies dedicated to RM, such as On Demand Manufacturing (Camarillo, California), have investigated other aspects of the build process (such as powder re-use in SLS systems) as it relates to long-term, consistent manufacturing.

Therefore, the following points are clear:

1. Traditional 'RP' technologies can be successfully applied to real-world manufacturing requirements.
2. Qualification of such technologies is, currently, the most important activity in RM, and will be required if RM is to be implemented industry wide.

## 14.9 Case Studies

Since the aeronautical industry tends to be held by very strict non-disclosure agreements, case studies of actual RM applications are very difficult to publish at this time. While the chapter should have made it clear that RM is at the heart of some very important aeronautical programs, full publication of the details of these programs will need to be left for a later update.

## Reference

1. Northrop Grumman (2003) Manufacturing technology: accelerating the factory of the future, SME WESTEC 2003 Conference, Los Angeles, California, 24–27 March 2003.

# 15

# Aeronautical Case Studies Using Rapid Manufacture

John Wooten
*On-Demand Manufacturing*

## 15.1 Introduction

The introduction of a new technology always involves either real or perceived risks (or both) to the end product. The benefits for implementing new technologies clearly must outweigh those risks in order for the change to happen successfully. This is especially true for aerospace hardware where certification of a new manufacturing process must be demonstrated and approved, and even then there are still concerns because of the potential consequences. Hardware is being manufactured today for the F/A-18E/F Super Hornet (see Figure 15.1) and other aircraft using technology that until recently was only used to manufacture prototype hardware. The benefits that allowed this technology to be implemented are discussed and compared to the risks that had to be overcome.

## 15.2 Problem and Proposed Solution

In February 2000, the US Navy partnered with the Boeing Company with the goal to reduce significantly the unit cost of the forward fuselage of the F/A-18 and begin the incorporation of six new avionics systems. In addition to reducing the unit cost, two other key goals were established: shorten cycle

*Rapid Manufacturing: An Industrial Revolution for the Digital Age*
Editors N. Hopkinson, R.J.M. Hague and P.M. Dickens 

**Figure 15.1** The F/A-18E/F Super Hornet. (Reproduced by permission of Boeing)

time from 34 to 18 months and reduce quality defects by 90 %. With the addition of the six new avionic systems, additional cooling was required. Also, there were associated components, such as electrical covers and shrouds, that would be needed to help meet these aggressive goals. Given the limited space in the jet fighter, cooling ducts (Environmental Control System, or ECS) needed to be designed such that the ducting could be accommodated in the space restriction while simultaneously meeting the goals. For several reasons discussed below, nylon-11 parts, fabricated by selective laser sintering (SLS), were the ideal fabrication solution if all the performance and manufacturing requirements could be met. Geometries of typical SLS-manufactured ducting showing complexity including integral attach features are presented in Figure 15.2.

Even though Boeing had demonstrated the feasibility to Rapid Manufacture components from SLS nylon-11, two issues needed to be addressed and resolved before this technology could be implemented:

1. ECS ducts are not primary structural components, but they must be engineered and have certain 'guaranteed minimum' mechanical and physical properties.
2. They must be made accurately enough to fit on to the platform and interface with the rest of the plane.

If these two conditions could be satisfied, then SLS could be considered as a viable fabrication method for the ECS duct. Because SLS would allow the ECS ducts to be built 'one layer' at a time, it would be possible to add in features that would have to be added or accommodated with additional parts if alternative fabrication technologies were utilized. Thus, the benefits started to become clear.

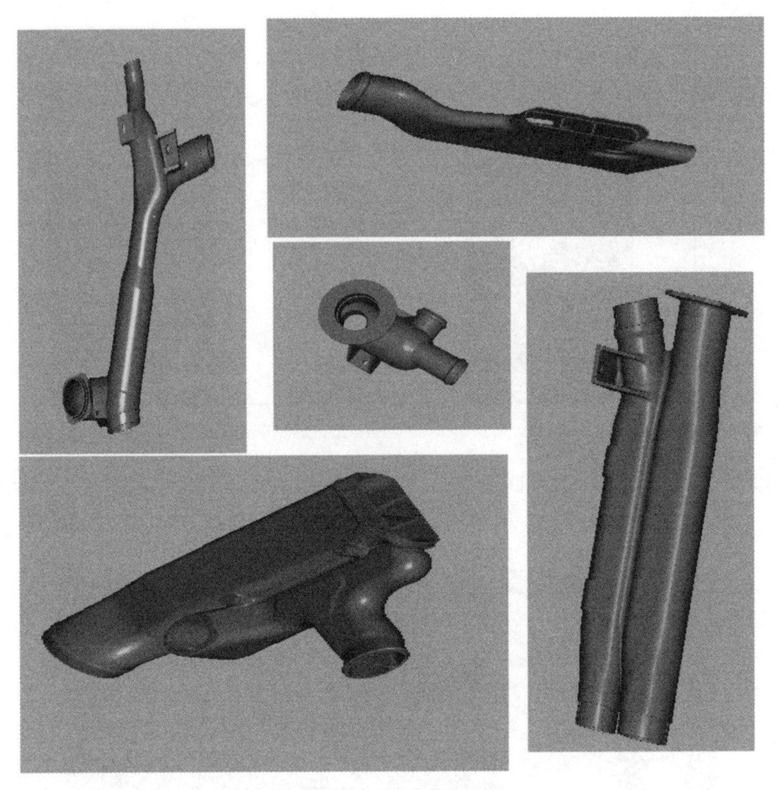

**Figure 15.2** Computer aided design (CAD) files for complex ducting components

## 15.3 Benefits of a Rapid Manufacture Solution

### *15.3.1 Design Flexibility Benefits*

The design flexibility provided by a layer-build process enables complex, ECS duct configurations to be manufactured that would be difficult or impossible to make by other fabrication technology, such as rotomolding or hand laid-up composites. These alternative techniques could produce parts with the same functionality, but it would require several individual parts that would have to be integrated with each other to produce the same function as one SLS part. Thus, the flexibility offered by SLS enables part count reduction, which helped the F/A-18 program meet its goals. An additional feature, which could be easily incorporated into the part, was the integral attachments. These are built into the ducts and eliminate the need for the standard or normal types of attachment mechanisms, e.g. hose clamps, and simplify the installation procedure and shorten the assembly time. Taking these factors into account, a system level cost analysis was performed and compared to ducting systems produced by rotomolding. As

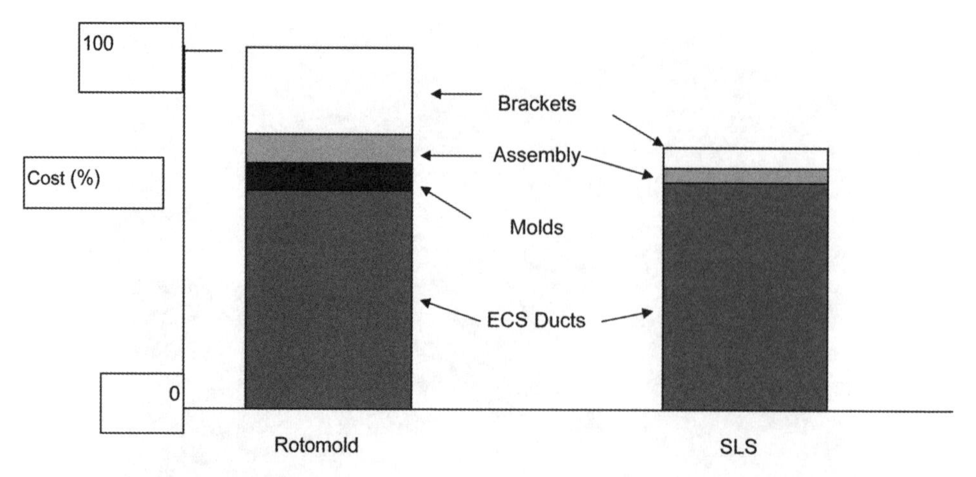


**Figure 15.3** Cost comparison between Rapid Manufacturing (RM) and rotomolding

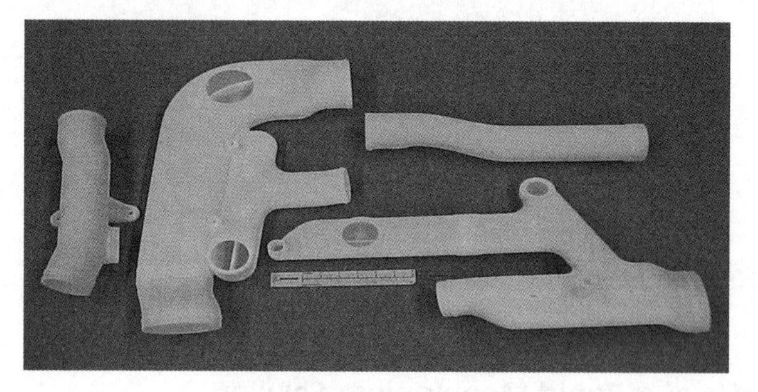

**Figure 15.4** Various components built as single units with RM

can be seen in Figure 15.3, approximately a 25 % systems level cost saving is realized.

Figure 15.4 shows various features, such as flow straightners, attach points and other integral, complex features, that are built into the parts to enable these cost savings to be realized. One indirect benefit of this technology was that preparation time on the plane for attaching the ducts is reduced; e.g. fewer holes were needed.

### *15.3.2 'No Tooling' Benefits*

Since SLS builds parts directly from CAD files, no tooling is needed. In addition to eliminating the cost of manufacturing and certifying the tooling, this plays another important role in facilitating the engineering changes that inevitably happen after production commences.

As design changes occur during the course of a long production run, the effect on the cost of manufacturing an SLS part is negligible. It basically requires loading in the new CAD model and building the part; however, this is not the case where a part is manufactured by a process that uses tooling. In this case, either new tooling has to be made or at the very least reworked. Depending on when the design change is flowed down, there is an additional cost on inventory. Because of set-up and tear-down costs, a process that uses tooling, like injection molding, will run several parts at one time and then put them in the inventory. If the inventory cannot be reworked, this inventory may become scrap.

Since parts are built and delivered only when needed with SLS, it is not necessary to maintain a large inventory. The inventory reduction is a significant cost savings, because whether it is kept at the supplier or the end user, it is still a cost that must be accounted for. Further, when changes are made, the inventory becomes scrap (as discussed above) or, at the very least, the parts will have to be reworked.

### *15.3.3 Systems Benefits*

Because of the elimination of attachment mechanisms, the reduction in number of parts and the conformal design solutions, a weight savings is possible with the substitution of SLS. For any flying platforms, this is always a benefit. The F/A-18 determined this to be approximately a 20 % weight savings.

As mentioned above, all of the parts could be built without tooling. This made all of the parts (about 60 different details) available early in the program and made it possible to air-flow balance the entire system, which had not been possible before.

## 15.4 Pre-Production Program

With all of these benefits, it was still necessary to convince the designers and program managers that the goals of the program could be achieved without exceeding the allowable risks. In order to do this a pre-production program was necessary. This required both recurring effort to develop the designs and non-recurring effort to evaluate the materials and parts.

A pre-production (or qualification) program was laid out. It included fabricating a material that could meet or exceed the design requirements for the chosen components and preparing designs. Once these designs were converted to .STL (Standard Triangulation Language) files, the parts could be built and evaluated. This effort included not only fabrication of the hardware but also fabrication of sufficient mechanical and physical property

**Table 15.1** Tensile properties from laser sintered parts

| Build plane | UTS (ksi) typical | YTS (ksi) typical | % Elongation typical |
|---|---|---|---|
| *Z* plane | 7.2 | 4.0 | 22 |
| *XY* plane | 7.4 | 3.9 | 38 |

coupons to verify that the parts could survive the loads, environment and life requirements imposed on these ducts. Thus, a series of tests were conducted using these coupons to verify that the requirements could be achieved and to establish a database. A summary of room temperature tensile properties is presented in Table 15.1.

Test beds were designed and constructed using the ECS ducting. The ECS ducts on these test beds were then subjected to environments that imposed static, dynamic and thermal loadings to ensure that they would meet the requirements.

## 15.5 Production

After testing of all the mechanical and physical properties of both the coupons and the ECS ducts was completed, a specification controlling the process was required. This specification ensured that the quality of the ECS ducting would continue to meet or exceed the requirements for the platform. Prior to production, a manufacturing and quality plan had to be developed and implemented that, when coupled with the specification, would ensure that the parts would meet the requirements.

On-Demand Manufacturing (ODM), a wholly owned Boeing subsidiary, was established to manufacture and supply these ECS ducts. Incorporated in June 2002, it began fabricating and delivering production hardware in December 2002. One year later, ODM had manufactured and delivered over 2000 ECS ducts to the F/A-18 program as well as similar-type components for several other platforms. The installation of some of these ducts is shown in Figure 15.5.

As discussed above, a significant benefit is the ability to make changes to production without incurring significant non-recurring costs. Several design changes were made in the first year of production. Had these ducts been fabricated by any process requiring tooling, the tooling would have required re-work or scrapping. With layer-build technology allowing parts to be built directly from CAD files, all of these non-recurring costs have been eliminated.

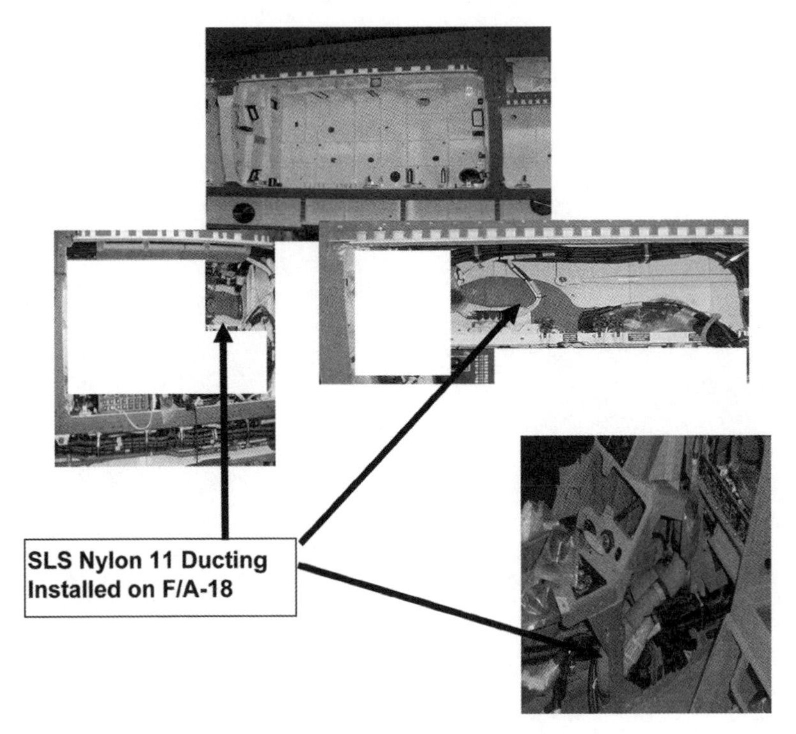


**Figure 15.5** Ducts in place on the aircraft

## 15.6 Summary

Layer-build technology was successfully implemented on the F/A-18 program to help achieve the goals established by the US Navy. Nylon-11 ECS ducting and similar components were designed and fabricated by SLS for a low-rate production application. The benefits, which included design flexibility, no tooling and just-in-time delivery, all contributed to the end customer accepting and allowing this technology to be implemented. The principal difference that allowed SLS to go from Rapid Prototyping to Rapid Manufacturing was the development of material properties that met the requirements and the establishment of a controlled process. Without an established material properties database, it would be extremely difficult to implement a new manufacturing technology.

# 16

# Space Applications

Roger Spielman
*Solid Concepts*

## 16.1 Introduction

Why can't we use this part? The question brought momentary silence around the conference table at Rocketdyne Power and Propulsion, a recent acquisition of The Boeing Company, which is world renowned for their expertise in the manufacture of liquid fuel rocket engines. The question was well timed, and we had no immediate answer to it. One of Rocketdyne's projects was the fabrication of the power system that was to provide the International Space Station with electrical power, and we had been busy designing subassemblies for that project. The current topic of discussion was a small enclosure that would house a capacitor measuring about 76 mm × 76 mm × 127 mm (see Figure 16.1). It would become part of a stacked array and integral to the system. The design called for an injection molded thermoplastic part, and the tooling for the part was already in the manufacturing cycle, expected to take another 2 months. The Rocketdyne Rapid Prototyping Center had been supporting the various Rocketdyne programs by providing rapid prototype models created using the selective laser sintering (SLS) process. The part that had generated the question was a capacitor box fabricated using glass reinforced nylon typical of the early–mid 1990s timeframe. As the SLS process was still relatively new at that point, both the materials used and the process itself was unknown for anything other than providing models or polycarbonate casting patterns. The nylon composite material was indeed very tough and fairly sound dimensionally. Maybe we should consider taking a look – it was certainly an attractive alternative to waiting 2 months for the completion of the injection

*Rapid Manufacturing: An Industrial Revolution for the Digital Age*
Editors N. Hopkinson, R.J.M. Hague and P.M. Dickens 

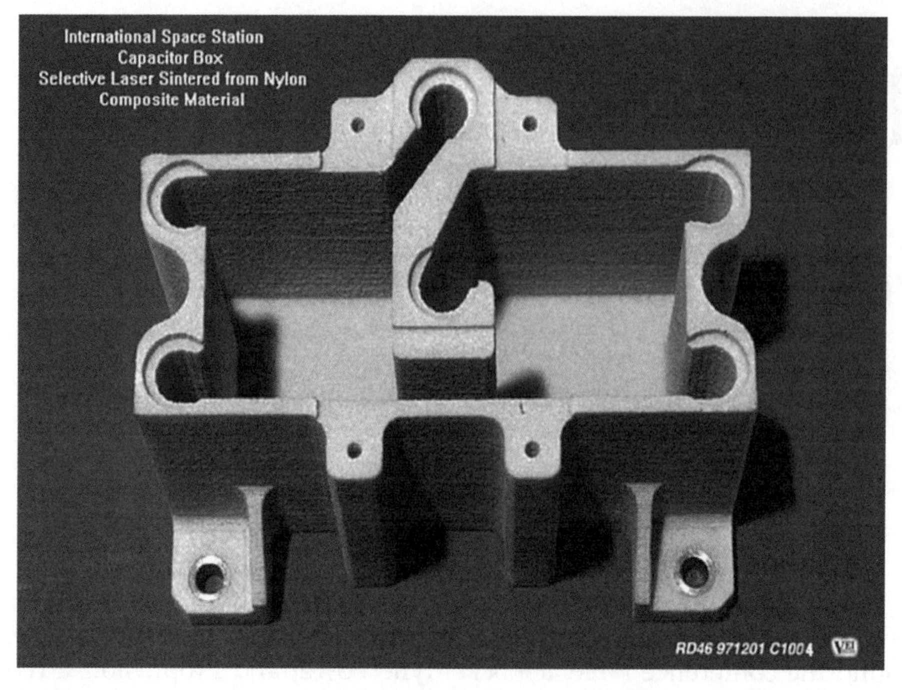


**Figure 16.1** Capacitor box for the International Space Station. (Courtesy of Boeing)

molding tooling. After all, Rocketdyne had purchased the SLS machine with the intent of producing actual usable hardware. It was time to step up to the plate.

The Rocketdyne Material Engineering and Technology Group launched a feasibility study to determine if the process and material could fit into the strict functional and environmental requirements of the Space Station. By the next meeting with the Space Station project engineers, we had good news. The initial investigation showed promising results using the SLS process and nylon composite material. However, nothing can be accepted 'as advertised' in aerospace applications, and certainly not in a 'man rated' environment. A formal material characterization was initiated to see just how this new process would perform.

## 16.2 Building the Team

As with any aerospace parts program, a regimented qualification agenda was set up that included all affected disciplines including members of the Polymer Engineering Group, Manufacturing, Quality Assurance and, of course, us. We were part of the Materials Engineering and Technology Group. Other disciplines would be called in as necessary as this was a first attempt to bring a new manufacturing process into the mainstream aerospace environment, which does not happen very often.

To obtain flight certification, the necessary paper trail must be created. This ensures that the finished parts will maintain the integrity that was originally conceived by the engineering staff. All materials used to create this part must be certified with established properties. As no known properties existed for the current material, the material properties had to be determined, documented and verified. The nylon composite material was investigated for tensile strength and elongation at break, toxicity, flammability and fungal tests due to their proximity to human environments in space. Material specifications, material processing procedures and manufacturing operation records were written. The entire material certification process can take several months, a little more time than it would take for Rocketdyne to have the tooling for the injection molding process to be fabricated. We pursued the qualification tests, selling the idea on the merits of an alternate method of manufacture. With the competitive nature even found in aerospace those days, a back-up plan that can maintain critical schedules becomes imperative to the success of the operation.

To certify these parts, the aerospace community would use a process we refer to as 'point design certification', meaning that for a given program, we could design the part for a specific job, using a specific material. All qualifications would be directed at that use. Rocketdyne's costs to perform the one-time certification were similar to that of creating injection molding tools for just one part. In an effort to streamline this process, we started a parallel effort to qualify material and establish manufacturing techniques and parameters for the SLS process. In short, there would be a slight delay and similar costs up-front, no delay and much lower costs down the line.

Due to the typical overbooking of a single Sinterstation shop, Rocketdyne Rapid Prototype personnel found it beneficial to capitalize on the assets available to a company the size of Boeing. With a phone call, the very helpful engineering staff at Boeing Commercial Air Group in Seattle, Washington, opened their doors (and two of their three Sinterstations) for use on our project. With both Rocketdyne and Boeing Commercial working the project, the required parts were completed about the same time the intended injection molds were ready for use.

As a side note, by the inclusion of Boeing Commercial Air Group on this project, an extremely important (and very beneficial) event took place – synergy. With the recent acquisition of the Rockwell International Aerospace business units (including Rocketdyne) by The Boeing Company, there was a very high degree of conjecture as to how the two aerospace giants would work together. The synergy created by this project between the former competitors captured managements' attention in a big way. This was something that every level of management associated (even remotely) with the project could report on. The end effect was enormous visibility and support from the management hierarchy.

## 16.3 Quality Assurance

By far, one of the most advantageous decisions made early on was the inclusion of quality personnel at the onset of the project. On top of attempting to qualify a new process (especially one as unfamiliar as SLS), we had the task of determining what dimensional controls were actually necessary. These parts were nested deep in an electrical assembly and dimensional accuracy was critical for assembly and operation. In the fabrication and assembly of rocket engines, dimensional control and tolerance stack-up is brought to a fine art. To introduce a new process that has limited control of dimensional stability is a tough sell. To bring the SLS process into a world accustomed to computer numerically controlled (CNC) tolerances would require an open mind and a shift in the typical engineering paradigm. What is actually necessary and what can we gain by allowing a relaxation in the typical four (or five)-decimal control? To understand process capability, the QA (quality assurance) personnel must have at the very least a working knowledge of the process – what affects the process and how those events will affect part quality. Bringing in the QA inspector early has many benefits. Encourage them to learn the process and understand the advantages and disadvantages of additive freeform fabrication. Let them participate in the design of experiments and design of any tooling or check fixtures, and by all means make sure they understand the financial and schedule benefits of using a process that requires no hard tooling, no lead time for the manufacture or storage for that tooling and the ability to revise the component design overnight.

In the case of our capacitor box, the final operation was tied into the inspection tool. As the design called for mounting these boxes with four integral mounting holes, we used undersize (pilot) holes, which were fabricated along with the part during the SLS operation. The drill fixture then located on the parts, providing not only correct location and sizing of the holes but also a check fixture for QA. Once reamed to size, a stainless steel sleeve was pressed and bonded into the part, making for a stable mount. As dimensional control was quite repeatable in the $X$ and $Y$ axes during build, the parts were oriented for best performance during sintering operations, which fashioned the weakest control orientation (the $Z$ axis, due to the inherent shrinkage of the nylon-11 material of the time) as the easiest to control. Additional 'machine stock' was added to the base of the part and fabricated during the SLS operation. This presented the opportunity to set the overall height (a critical dimension) through a simple (single) machine

**Table 16.1** Typical dimensional data recorded for each build

| | Actual dimensions (inches) Machine #2 Run date: 4-23-97 | | |
|---|---|---|---|
| Part number | Length | Width | Inside height |
| 1 | 4.143 | 2.928 | 2.354 |
| 2 | 4.139 | 2.931 | 2.351 |
| 3 | 4.143 | 2.924 | 2.352 |
| 4 | 4.144 | 2.927 | 2.352 |
| 5 | 4.144 | 2.921 | 2.355 |
| 6 | 4.144 | 2.927 | 2.355 |
| 7 | 4.140 | 2.922 | 2.356 |
| 8 | 4.141 | 2.925 | 2.357 |
| 9 | 4.144 | 2.922 | 2.366 |
| 10 | 4.144 | 2.926 | 2.362 |
| Average | 4.1426 | 2.9253 | 2.356 |
| Drawing | 4.14 | 2.93 | 2.35 |
| Delta | 0.0026 | −0.0047 | 0.006 |

surfacing operation. All dimensions and material properties were recorded for each build (see Table 16.1). Sacrificial process control samples (tensile bars) were fabricated during each build and this information was used to qualify each build.

Quality of the powder material is also a critical factor in part strength. The stress group of engineering had determined the design allowables for the performance of these parts. To qualify, each batch of the material was subjected to acceptance criteria of its own before ever becoming a measurable part. New powder batch lots were qualified in accordance to color of material, uniformity and sieve analysis. These operations were performed before the incoming powder was loaded into an SLS machine, and then only for qualification of mechanical properties. Once the powder passed these incoming inspection criteria, it was re-marked 'Flight Material Only' and was segregated from the other powder materials used for general prototype or casting pattern use. All of this acceptance criteria was spelled out in the new SLS material specification – the first as far as I know for any additive freeform fabrication process that was 'Flight Certified'.

## 16.4 How to 'Qualify' a Part Created Using This Process

The SLS machine was qualified first. Specifications were written to establish baseline parameters and processing procedures were written to establish cleanliness standards. These operations were to be routinely performed on the machine to ensure that it was in top form and capable of repeatable control.

Material properties were monitored run by run, and this was the determining factor in 'selling' the completed SLS build. Each 'build' was comprised of numerous parts as well as process control samples, which were fabricated in each run. The material properties as measured from these samples governed the most significant component of the acceptability criteria for the parts included during any particular build. Other criteria included dimensional control and the absence of any visual anomalies. This meant that the part had to be strong enough (proven by the process control samples), it had to fit correctly and it had to look right. Any voids, cracks, layer delaminations or other visual anomalies were cause for disqualification.

## 16.5 Producing Hardware

This is not quite as simple as loading the Sinterstation and starting the build. We now had an accepted process, and that means paperwork, checks and balances and manufacturing operations record (MOR) books. A MOR book is the legal document we use to fabricate hardware. It is a step-by-step set of directions with places throughout to assign a mechanics stamp or other form of identifying exactly who has performed the work, and whether or not they are qualified (and current) to perform any particular task. This record starts with accumulating everything you need to do your job. In our case, qualified powder was the first step. This ensures that you are not only using the correct material but that it has passed the rigorous acceptance criteria described above. Once the machine is proven to be clean to specification and loaded with powder (also proven acceptable to specification), we can load the build parameters using the proper (written) procedure and of course the part files or build, which is also set up in accordance with the SLS material processing procedure (MPP). The paper trail has begun, linking the final part back to the raw material it was produced from.

An interesting point was that these initial operations were performed by an engineering entity rather than a manufacturing group. As this was a new process and in the early stages of acceptance, this was the most expeditious means to produce the parts. This also created some interesting situations. Even though we brought in the raw material (nylon powder), as soon as it came out of the Sinterstation it was a part. This meant we could do the post-processing as required as long as it did not leave our room. Moving hardware at Rocketdyne was the function of a different bargaining unit (commonly called a union). Moving an SLS flight part out of our little

**Figure 16.2** Retainers for the International Space Station

domain without utilizing union people would result in the filing of a grievance. This certainly puts a different twist on 'rapid'. Therefore all post-processing of these parts was accomplished in the Rapid Prototype Center. Inspection personnel were brought in to do their acceptance verification, with the final packaging (per specification) operation being the last we could do before calling the expediter to move the hardware.

The capacitor box project was deemed successful and several other parts for the Space Station were channeled to us for manufacture using the SLS process. One part in particular locked in the benefits of using digital information and layered manufacturing. While testing components at Cape Kennedy a potential problem had surfaced. During vibrational testing – subjecting the assemblies to launch loads – a ribbon cable had vibrated loose on an electrical assembly. Project designers came up with a simple retainer design to eliminate this problem (see Figure 16.2), but time was critical. There was no time for 'conventional' manufacturing (injection molding), so they turned to us. We had performed well for several other parts, how could we perform on this one?

Two hundred of these retainer brackets were required, and they only had weeks, not months to complete the retrofit without incurring any launch delays. A design was forwarded to the Rocketdyne Rapid Prototype Center for evaluation, and it was found that with some creative nesting of the parts, all two hundred could be fabricated in one run. It should be remembered that this was done using a DTM Sinterstation 2000, which utilized a 12 inch (305 mm) diameter by 16 inch (406 mm) deep work envelope. The ancillary operations (measurement, reaming, insert bonding and part number application) would need to be streamlined as well. As we were already in 'production' mode, a very aggressive SLS build (for that time) was set up and started, following the established policies and procedures.

While setting up the initial build, Quality was brought in ahead of time to assist in streamlining the process. We found a major schedule driver to be part identification. The special inks used for stamping the part required a union mechanic with special certifications. The time estimate for this operation was almost a week in itself for 200 parts including the 24 hour ink drying period. We proposed to Quality that since all of the parts were the same part number (which was to be ink stamped on each part), we could add the raised part number to the file and build it into the part – similar to the identification on a casting. This was agreeable to our Quality representative as the discreet serialization would occur only on an attached tag. This move alone saved us a week in schedule.

The build would take almost two days, so while the machine was running, the 'team' was pulled together. This is where you really benefit from having the appropriate disciplines up to speed concerning the technology. The team set up the schedule of events to take place for the other ancillary operations. These would be handled this same way as the capacitor boxes, with Quality personnel setting up shop in our area in order to perform the dimensional and visual inspections, bag, tag and serialize the parts. Everybody had their task clearly identified and were ready on completion of the SLS operation.

When the smoke cleared, we had fabricated, performed all post-build operations (reaming mount holes and inserting the metal sleeves required by the design), full acceptance and certification for flight status, packaged and shipped the full compliment of parts in one week – unheard of in the aerospace community. SLS and rapid manufacturing were truly on their way as an accepted process.

# 17

# Additive Manufacturing Technologies for the Construction Industry

Rupert Soar
*Loughborough University*

## 17.1 Introduction

This chapter describes the application of additive manufacturing technologies (AMTs) to the construction industry. It is intended for both readers with construction, civil and building engineering and architectural interests as well as the Rapid Manufacturing (RM) pioneers. It will build on, and assume the reader is familiar with, the background AMT terminology and processes covered in the earlier stages of the book. Many of the aspects discussed are not intended solely for the construction industry but are applicable to many large-scale freeform fabrication applications. The choice to focus on construction is based on progress made in developing this field of research and its application.

For readers of a construction or architectural leaning, this chapter aims to disperse a misconception which has emerged within the last few years. A search of AMTs, applied to the generic field of construction, will turn up examples of the technology being used for little more than scale model making. This chapter will not be considering these aspects.

Within the construction industry, AMT development is underway. When referring to AMTs in this application, a degree of caution should be applied. By associating the term with construction, there is a danger of assuming that no clear examples of AMTs exist or existed, prior to the last few years. A

*Rapid Manufacturing: An Industrial Revolution for the Digital Age*
Editors N. Hopkinson, R.J.M. Hague and P.M. Dickens 

straight comparison of conventional construction and existing Rapid Prototyping (RP) techniques produces interesting parallels. Current RP and emerging AMT techniques may be characterised as having the following attributes:

- Material preparation
- Material delivery
- Layer-by-layer fabrication
- Phase change of either a build or bonding medium
- Support system/support removal
- Finishing/post-processing

Relating these attributes to general construction practice, it is clear that there is a broad similarity between the two. Traditional construction is analogous to a large build chamber where both manual or automated material preparation and delivery systems build in a layer-by-layer approach. Supports are incorporated through scaffolding systems and finishing/post-processing relates to decorative finishes and second fixings. In fact, in some ways, conventional construction attains what AMTs are striving for in that they already utilise:

- A multitude of materials into a single structure as per functionally graded structures
- Differentiation or 'mass customisation' in numbers from 1 to 1000
- Embedded or integral circuitry and distribution/service networks on a par with emerging 'direct writing' techniques
- Construction at multiple scales simultaneously, ranging from fabrication of the 'macrostructure' down to placement of individual tiles and fixtures at the 'microlevel', resembling 'hierarchical' structures
- High levels of design and structural optimisation in terms of weight reduction, strength, thermal, acoustic, ventilation, etc.

Therefore, so if construction is this advanced then why do we need to consider AMT's solutions? Within construction, the levels of complexity, described above, are accomplished manually or, where automation does exist, it tends to be automation of a single task and not a multitude of tasks simultaneously.

## 17.2 The Emergence of Freeform Construction

### *17.2.1 Applying Lessons from Rapid Manufacturing*

Specific drivers for change, in construction, are numerous and well documented and fall outside this chapter (Gibb, 2001). However, many of the mechanisms leading to the need for technological solutions in construction can be seen in modular technologies, such as FutureHome (Wing and Atkin, 2002) and the Japanese skyscraper layer manufacturing systems (Howe, 2000), neither of which are driven necessarily by cost reductions and time constraints alone.

By way of background, and a short detour from our theme, the FutureHome project strove to integrate higher levels of design freedom within a factory-driven modular approach, with the production of the completed structure being divided between automated fixed production units and 'field factories' (Wing and Atkin, 2002). Modules would be manufactured off-site and transported for assembly on-site. Essentially, the project aimed to strike a balance between the advantage of moving fully assembled steel frame modules, over large distances, while utilising local construction expertise and robotics for the final assembly. The project, therefore, focused on:

1. Retraining of on-site staff for dry construction techniques.
2. Increased on-site automation, through robotic gantry crane systems for handling and placing modules.
3. Design of flexible module interconnects for wiring, waste and service ducting.

FutureHome may well become a template for both on-site and off-site construction methods in the near future. In terms of the scope and range of issues it addressed, it may define the next logical step to meet the environmental and physical challenges facing the industry.

Towards the latter half of the 1980s, the exact dates are difficult to pinpoint, significant developments took place in Japan to fundamentally change the approach by which large skyscrapers were constructed. Motivation for new techniques varied but key considerations were the hazards faced by skyscraper construction workers in a region of the world known for high winds and earthquakes.

Though solutions varied, all attempted to create an environment, on site, where either workers or automated systems could work under cover, for which Howe (2000) categorises the methods into three types:

1. Collections of function-specific robots that work independently of each other and are not laid out in a systematic way.
2. Robotic systems that form a systematic factory, fixed in the context of the site.
3. Robotic systems that form a systematic factory capable of moving itself along.

Within the first category, Howe places the numerous examples of individual robots developed for the construction industry. These range from autonomous or manually controlled concrete finishing or trowelling devices up to, for example, autonomous devices for the assembly of tiles to external facades or window cleaning devices. Within the second category are examples of skyscraper construction which 'extrude' each layer of the structure skywards. Two companies to exploit this approach were Kajima Corporation, with a technique known as the 'Automated Building Construction System' (AMURAD Construction System), and Fujita Corporation's 'Force up Building Floors High into the Air' (ARROW-UP System). In both, a purpose-built automated factory was erected on what was the first floor. Materials, components and even modules were fed into the automated system, which would assemble the entire first storey. Once complete, a series of jacks would elevate the first storey upwards, clearing the assembled structure from the ground, to allow the construction of the second floor beneath it. This process was repeated with each subsequent storey being forced upwards beneath the others. Once the last story was complete, the assembly plant would be cleared from the ground floor and the structure finished. Howe cites two high-rise structures completed using this process in Nagoya and Tokyo. To draw an analogy with this technique to current RP processes, EnvisionTec's Perfactory$^{TM}$ process is a suitable candidate for 'top layer down' build strategies.

Most RP aficionados, however, are more familiar with RP systems that build progressively by adding layers to the top of the structure or 'bottom layer up' techniques and, in terms of skyscraper construction, these methods fall into Howe's third category. At its peak, there were a further six companies building structures throughout the Western Pacific Rim and it was Shimizu Corporation who coined the term 'Layer Manufacturing'. Shimizu Corporation developed its SMART system, Taisei Corporation's T-Up, Takenaka Corporation's 'roof push up method' and Obayashi Corporation's ABCs system, to name just four. The common denominator for all these methods was that a large hydraulic canopy was erected over the uppermost storey of the structure to allow the construction of a single storey. Once complete the canopy would be jacked up so that the next story could be fabricated. Though different methods were utilised, for each method the intention was to integrate high levels of automation and assembly robotics, within the canopy, so that construction could progress unhindered, in changing weather conditions, with little or no risk to human operators who were there to oversee the process. Different strategies were employed to lift the assembly plant by one storey at a time; for example, Shimizu's SMART and Taisei's T-up systems built the structure around a rigid central core from which the factory could be jacked up, whereas Obayashi's Big Canopy placed steel columns at each corner of the structure.

These are, without doubt, remarkable scales at which a layer-by-layer approach can be taken and it may come as a surprise to know that, without exception, none of the techniques are in operation today, even though all of the companies in question have details of the methods listed on their active websites. When asked why these methods are currently not in use, Mr Shigehiko Tanabe, director of the Construction Engineering Group Institute of Technology for Shimizu Corporation, was naturally only able to speak about Shimizu Corporation's reasons, but stated that, in addition to the downturn in the global economy, there were two additional reasons relating to the process itself. The first related to the space in which the assembly factory must be sited. Skyscrapers tend to be placed within limited space constraints, normally surrounded by other skyscrapers. This makes it difficult to feed a SMART factory with materials as construction progresses. In many cases an entire supply and pre-assembly plant is required on one side of the building to keep production going. The second reason related to form. Increasingly there is an expectation, by clients, that larger tower blocks are no longer cubic. With an increasing demand for non-cubic structures there is the problem of where to position the elevating columns so that they support the assembly platform uniformly. However, Howe (personal communication, 2005) points out that, even though layer methods are not in current use, this condition should be temporary and is expected to pick up in the near future.

Though FutureHome and Layer Manufacture represent a modular approach they also fall under a new class of emerging construction methods, becoming known as 'Digital Fabrication'. Digital Fabrication techniques essentially encompass 'the CAD/CAM driven manufacture of 'dry' sub-assembly components, formed via subtractive or formative processes' and interestingly begin to introduce the concept of 'freeform' to construction. Some of the best examples can be seen in Foster and Partner's Swiss Re tower and Gehry & Associate's zolhoff towers, amongst others.

To get back to the subject, though the construction industry has been accused in the past, it currently cannot be accused of resisting the change from traditional methods and many of the underlying causes are paralleled by those changes that led to the emergence of AMTs, such as:

- Automation
- Diminishing skilled labour
- Synchronisation of the production process
- Reduction in part count
- Greater material utilisation

Though both Rapid Prototyping (RP) and Rapid Tooling (RT) solutions were indeed driven by cost and time constraints, current and emerging AMT solutions are increasingly being considered for the value they can add to a product in terms of function, aesthetic value, etc., which may exceed the capabilities of their predecessor. In line with the expanding application of AMTs into all aspects of human endeavour, it is therefore not surprising that researchers began to ask whether AMTs could be applied to construction. Taken that the Japanese Layer Manufacturing systems prove that AMTs, in the broadest sense of the definition, can be applied at the extremely large scale, then what about the intermediate scales, covering the vast majority of civil structures, from the humble garden shed to the office block.

In 1997 Joseph Pegna grappled with this question by considering both the constraints and issues associated with possible larger-scale AMT solutions (Pegna, 1997). Pegna considered the problem in terms of scale and the specific issues associated with incremental increases in scale. He began by classifying AMT solutions (available then) in terms of material delivery and deposition rates, and then calculated typical material delivery rates for an average two-storey house of height 7.5 m (not including the roof), a base frame of 200 $m^2$ (approximately 15 m by 14 m) and a total construction time of 1440 hours (assuming 2 months at 24 hours per day continuous operation, or 3 months with 33 % down time of the machine). This gives a required volumetric flow rate of 1.04 $m^3$ $h^{-1}$. This simple calculation readily demonstrates that if any AMT solution is to match existing construction production rates then processes such as laser phase change and inkjet technologies would be out of the running (this would need to be re-addressed today with some of the new high-speed sintering processes under development). To find a solution, for a process to meet these deposition requirements he turned back to the construction industry for inspiration for a selective aggregation process.

The solution was both neat and innovative and he proceeded, pretty much entirely self-funded, to design, test and validate a blanket sand deposition process followed by selective deposition of Portland cement through a mask, using water as the binding agent. The application of water to sand and cement does not produce a robust structure, so Pegna developed a pressurised steaming process which proved sufficient to generate structurally robust parts for testing. One of the interesting outcomes from the research, and one that the author stressed needed more investigation, was that resistance to tensile loading was unusually high for the concrete specimens produced, with higher achievable packing densities. He went on to highlight the potential savings in using and recycling unused material back into the process by utilising filler materials from recyclable sources, such as clinker or power station gypsum, for example.

Importantly, Pegna introduced a departure from current thinking. Much of the current 'state of the art' in construction, epitomised by the Japanese Layer Manufacture and FutureHome modular methods of construction, relate to the fabrication of subassemblies, analogous to aircraft and ship manufacture, in that individual components are manufactured using existing subtractive and formative techniques (as defined by Burns, 1993). These are then assembled to make the whole. The reasons for this approach are numerous but the attraction lies in the multitude of components/materials that may be assembled in a limited space, clear of milling and moulding machinery, thereby generating waste off-site while reducing waste, injury, etc., and maintaining access on-site. In traditional construction techniques the opposite is the norm. Cutting, machining, mixing of aggregates and plasters, as well as moulds for concrete forming are performed directly in the assembly area (or at least close to it), and these have tended to become known as the 'dirty or wet trades'. Specifically the construction industry now draws a distinction between 'dry construction' and 'wet construction' methods, with much of the emphasis leaning towards the elimination of 'wet construction' methods. One of the key aspects, which AMTs will introduce to construction practice, is that there are greater benefits to be gained through computer-controlled preparation, delivery and deposition of wet or wetted liquids, powders, binders and aggregate systems than purely dry systems. These will become a 'third way' and are, appropriately, being termed 'freeform construction' solutions.

### *17.2.2 Opportunities for Freeform Construction*

Before we move on to discuss actual solutions, it is worth re-emphasising the fact that, if AMT methods are to find a place within construction, they will be competing in one of the most demanding cost and time driven industries in the world. However, as with the emergence of AMTs for manufacturing, medicine, pharmaceuticals, automotive, aerospace and consumer goods, there is 'added value', through greater functionality, to consider. If this is the case then it is worth considering what AMTs for construction may be able to offer over and above current capabilities.

#### Geometric Freedom

Much of what the public currently perceive as quality, in 'fashionable' consumer-driven products, can be related to three-dimensional computer aided design (CAD) capabilities. This perception differs from traditional views on quality, where the public had an, arguably inherent, association between an artefact's complexity and/or uniqueness in relation to its cost and/or status. AMTs will inevitably enhance the changes which CAD design has begun by making complex, exotic or organic forms available to mass markets. This has a consequent effect on the public's expectations for buildings. Without exception, any major civil construction project will include a freeform aspect to the design if it is expected to make an impact. However, this is where the influence seems to end. Buildings or structures that fall below the multi-million dollar remit tend to fall back towards cubic designs. Modularisation works in opposition to freeform design and it is this aspect that freeform construction will have a visible impact on first.

By far the largest building sector is for civil housing, offices and shops. Unfortunately, construction design for mass markets rarely inspires major product launches, as would be expected for a new consumer product. Form is derived from the traditional requirement to lay bricks or blocks in straight lines to minimise construction time, and freeform geometries using traditional construction methods cost more than the average citizen can afford (Tsui, 2002). Thus it takes a small leap of imagination to envisage the effect that geometric freedom may have in the way the public perceive and purchase mass civil construction. In terms of marketing, freedom of design may well herald designer housing which may, in turn, drive fashion housing, which may, in turn, drive a market to demolish and re-build regularly, which may well solve one of the greatest problems, facing almost every government, in persuading us to upgrade our houses and offices in the light of changing climatic conditions and energy conservation. As Joseph Pegna observed, if the same materials can act as filler material for the new structure then a large grinding machine and a large solid freeform fabrication machine device may be all that is required on site.

## Integration of Form and Function

With few exceptions, most civil structures utilise the Froebel block (wooden building blocks) or cubic design approach. Cubic design is favoured as vertical walls are self-supporting and can be built in simple runs of bricks, closures for ceilings and openings, by beams and lintels, with little recourse to additional support mechanisms. As a design moves to the 'grander' scale, then arches, buttresses and domes are required, which normally incur the additional cost for support systems during construction. Arches, buttresses and domes allow the dissipation of loads and maintain stability within a structure more effectively and allow for construction on the massive scale. Pushing the limits of masonry construction, e.g. in cathedral design, was only possible through this approach. At the pinnacle of this approach was the self-supporting dome structure, such as Emperor Hadrian's Pantheon (Rome) or Brunelleschi's Cupola (Florence), where the structure was so large that support systems could not be considered. It is, therefore, no surprise that we tend to associate non-cubic designs with high art and extravagant design.

What has never been possible to consider, in construction terms, are the potential benefits of constructing structures for mass markets (i.e. residential, light commercial and commercial properties) using non-cubic or freeform design. If we accept that CAD systems can utilise non-cubic forms, then how can they be built? Here lies the second of the potential benefits which freeform construction should bring about. If indeed a building can be designed and built in this manner, then the structural elements can be combined with what we already perceive to be aesthetic elements, such as the arches, domes and buttress elements discussed. There are also secondary incidental benefits which include, theoretically, the elimination of much of the heavy lifting equipment used to place these structural elements. Removing this equipment improves security (as they are not there to steal) and produces lower risks to health through injury.

Relating this to current AMTs, the use of supports perform an identical role. What AMTs can learn from construction is the use of self-supporting

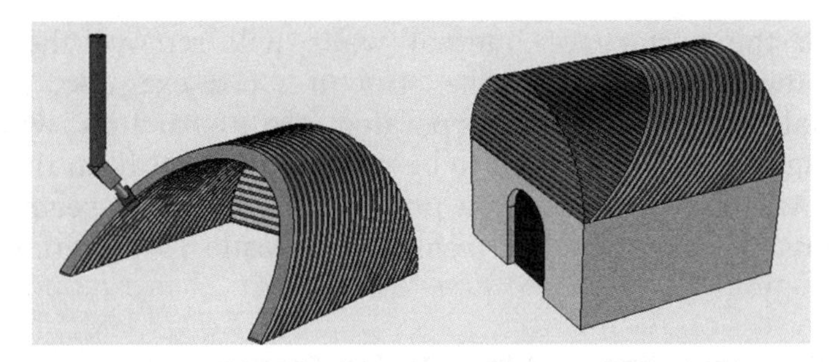

**Figure 17.1** Method for fabricating a self-supported vaulted ceiling. (Courtesy of Behrokh Khoshnevis, 2004)

non-linear forms derived within the CAD software during the actual design process, as shown in Figure 17.1.

## Integrated Services

Although opportunities lie in design freedom of the external structure, there are also significant opportunities for the internal structure. The internal structure can include either the habitable space within a dwelling or what lies within the walls. In the first instance, there may be opportunities to fabricate fixed items, such as customised shelving, storage, fireplaces, dado and ceiling roses, etc., during the fabrication phase. In the second instance, opportunities exist to integrate current 'conformal channelling' systems for internal detail such as ventilation and trunking systems, as shown in Figure 17.2(A), as well as building specific service trunking containing, services, heating ventilation and air conditioning (HVAC) and wiring systems which click together within the structure, as demonstrated in Figure 17.2(B), and will, in time, be extended to include the 'direct writing' of both electrical and optical systems.

Integration of these systems is not that far away. Moves to low-voltage distribution networks and ultimately to optical fibre networks embedded

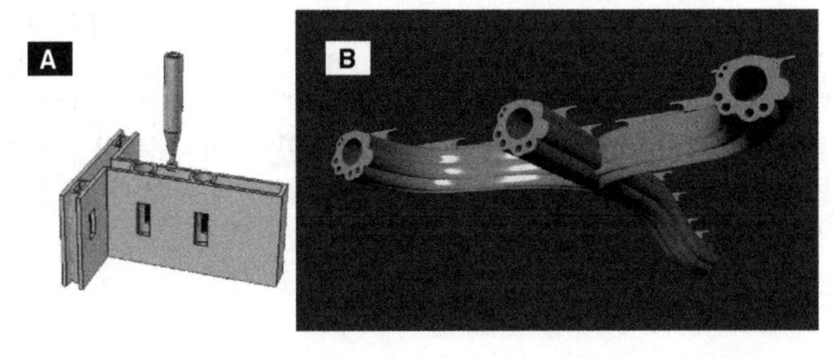


**Figure 17.2** Examples of integrated services and service modules built by AMT systems. (Courtesy of Behrokh Khoshnevis, 2004(A) and Rupert Soar(B))

just beneath the surface of internal walls will remove the danger of integrating these systems into the structure. For example, simple low-voltage metal grid networks, incorporated into a structure, will allow for light emitting diode (LED) lights to be pushed into a wall in the same way that thumbtacks are pushed into a pin board. What may seem impossible today is possible tomorrow as technologies jostle to accommodate each other.

## 'Black Box' Capabilities and the 'Design Freeze'

Great effort is being expended to make the link between the rigidity of parametric design and artistic flare. Haptic and visualisation systems are beginning to break down the barriers to the point that non-designers can interact with and modify a design. This will be possible where the software calculates the consequences of a design intent in terms of the consequences on the structure, through techniques loosely termed 'black box' systems. Black box systems are invisible to the person creating a design and interpret the design intent in terms of the process required to produce the object. This is currently being used with computer numerically controlled (CNC) machining applications such as those developed by TNO in Holland. The implications for freeform construction are significant in that the person designing a structure, and this may ultimately include the customer themselves, can explore many iterations of a design while the underlying software is calculating the implications of that intent. Parameters for process capabilities, material characteristics, structural requirements, space, services and HVAC requirements would be embedded within the program so that a design which contravenes an optimum state will be highlighted to the designer, with corrections suggested, during the design process.

This level of interaction into the design process will also require the inclusion of a 'design freeze'. High levels of interaction, with the ability to see the effects of a design instantly, will inevitably lead to more iterations of a design being considered, not unlike the role the architect plays in interpreting a client's requirements into the final design. Implementing a design freeze in the design process is analogous to pressing the 'print' icon. Once the printer is committed to printing that page it cannot be stopped and, in the same way, this will be true for freeform construction. The act of physically pressing a print button will initiate the build sequence, from which it is physically impossible to stop or change the design again without incurring the cost of going back to the design phase.

## Structural Functionality

As with functionally graded materials (see Chapter 7), freeform construction processes will enable the structure to be optimised, in terms its functional requirements, through the selective deposition and grading of different materials. Structural load dissipation is probably the most obvious application, with strengthening materials being deposited simultaneously with a more general construction material, as defined and derived from both finite element analysis (FEA) and computational fluid dynamics (CFD) analysis of the three-dimensional CAD form. Though not obvious, there is interest in this capability, the work being conducted by Loughborough University, from the military, who were quick to realise the potential of selectively incorporating low observability (LO) rendering particulate into fabricated bunker structures.

## Structural Optimisation

Structural optimisation will offer the construction industry significant capabilities over and above all previous and current construction processes. Optimisation of a system containing multiple constraints/variables assumes that there is no single solution that satisfies all conditions. Optimising a system implies that there are different solutions that satisfy all constraints to a greater or lesser degree. Which solution dominates depends on the state of the system at that point in time. AMTs potentially offer the first approach by which optimised structures can be derived, within a CAD system, and be reproduced faithfully into a physical structure or component. Initially, freeform construction will offer static solutions. In other words, once an optimised solution has been derived computationally, the physical manifestation of that structure cannot change within the physical environment in which it exists.

This book has previously covered the opportunities for AMT component optimisation (see Chapter 2). Research between Loughborough University, BPB plc (a construction systems specialist) and Z Corp is looking at optimised solutions for freeform construction. In particular, organic structures are being investigated in the hope of exploiting what may have one of the greatest impacts to the construction industry. Neglecting waste from excavation, the construction industry currently creates 70 million tonnes of waste per year in the UK alone, four times domestic waste production (Guthrie *et al.*, 1999). In particular, the waste stream generated is often mixed and, to some degree, hazardous. Impending environmental legislation will, and already is, forcing the industry to manage its waste. Figure 17.3(A) shows that a typical cross-section of a pre-assembled external wall, including fixtures and fittings, may contain around 15 different materials embedded within. Each of these materials generates waste, which is both mixed in the factory, prior to assembly on-site, or upon demolition.

AMTs will offer single material solutions and, to this end, the research is exploring both the process required, to produce such structures, as well as the possibilities of fabricating customised panel sections whose geometry

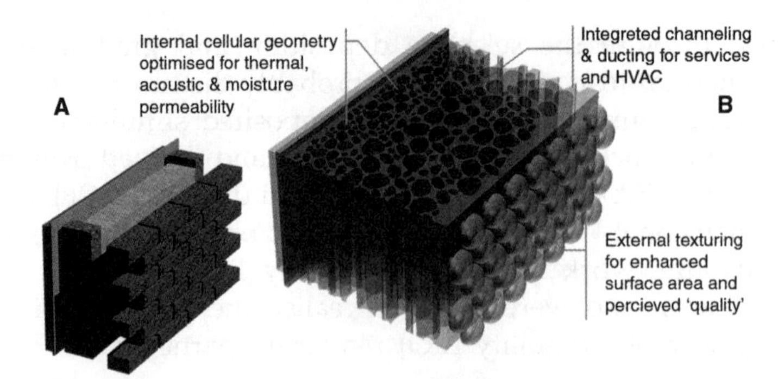


**Figure 17.3** (A) Typical cross-section through an external wall element and (B) impression of the capabilities for a single material external wall section.

reflects an optimum balance of structural, thermal, acoustic and ventilation properties, as the impressionistic image in Figure 17.3(B) demonstrates.

Initial work is focusing on geometric solutions to each of the main constraints, after which simulations will be run to identify optimum geometric solutions to all of the constraints simultaneously using 'morphing' simulation software. In addition, aspects such as variable and passively driven moisture permeability and movement through the structure are being explored by, for example, the incorporation of textures capable of increasing the surface area of the external surface to enhance or hinder moisture movement through and around the structure.

## Structural Homeostasis

Organic systems optimise dynamically, in that they possess the ability to physically change their form to remain at an optimum, within a set of fluctuating environmental constraints, that it encounters on a daily basis. Trees, plants and corral, like current AMT processes, are additive processes but, unlike AMTs, trees continue to find optimum solutions on a daily basis through growth. Animal structures, such as skeletal systems and the physical structures built by social insects, go one stage further by incorporating both additive and subtractive capabilities within the system. In the case of animals, this is a requirement of mobility, as over-engineering will hinder an animal from escaping a predator or capturing prey, and, in the case of social insects, to allow reinstatement of the system after damage or predation.

There is technically no limit to the level of complexity of geometry that can be integrated into a freeform structure. In line with this objective, researchers at Loughborough, Cambridge and Syracuse Universities, with collaboration from the Namibian National Museum, Namibian Ministry of Agriculture and BPB plc, are investigating the level of integration of passive environmental control mechanisms that can be incorporated into human habitats. Specifically, the research seeks to understand the method of passive control exhibited in structures built by social insects. It has long been known that some ant and termite structures possess, within them, adaptive capabilities that result in high levels of self-regulation (known as homeostasis) of the internal environment of the habitats they build. Bees and wasps are also social insects, but tend to achieve a homeostatic equilibrium by coordinating their activities, which includes simultaneous wing beating in a particular direction to create air currents that both regulate nest temperature and replenish respiratory gases deep within the colony.

Some of the higher orders of termites (Macrotermitinae) have gone one stage further by evolving freeform construction processes who's product are geometric forms capable of controlling the internal environment. Those familiar with these structures may be aware that some of these mounds utilise a central open chimney, sometimes rising many metres into the air. In the same way that our own chimneys work, buoyancy-driven respiratory gases are drawn up through the mound, which both reduces nest temperature and replenishes respiratory gases. There is, however, one particular family of termites who have evolved a construction method known as the 'cathedral' or closed structure. These tall closed structures, shown in Figures 17.4(A) to (C), attain incredible levels of environmental control without recourse to an open topped chimney. These structures are found throughout sub-Saharan Africa and are able to regulate internal environmental conditions by inducing or retarding energy flows both into and out of the structure, and probably represent the ultimate in 'smart' construction. Of particular interest is the fact that they achieve this without recourse to electromechanical controls, utilise only renewable energy sources, which naturally implies that they are not restrained as to where they can build (as we are).

Though these structures have been studied for over a century, there has never been the capability to understand how such levels of control can be achieved through, what appears to be, geometric rules. Modern three-dimensional scanning techniques, advanced simulation and freeform construction potentially

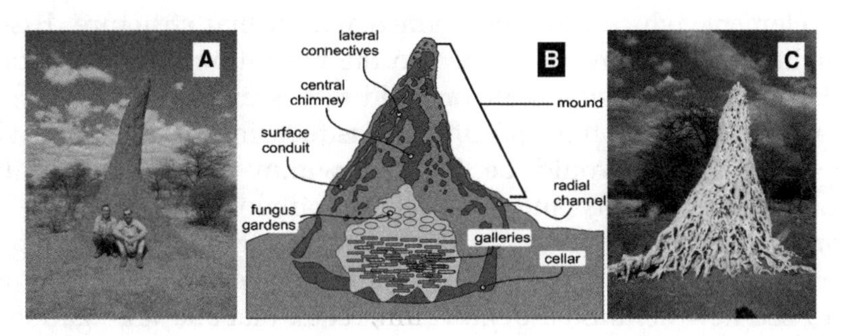


**Figure 17.4** *Macrotermes michaelseni* mound with a schematic and exposed underlying vascular system

represent the first time that humanity has had the chance to capture, model, interpret and translate these geometric rules into our own habitats.

This will not be an easy task, as Figures 17.4(A) to (C) demonstrate. What appears to be little more than a pile of earth (the lean is intentional as the structure is optimised for constraints such as rain driven erosion versus heat induced construction, as the sun's zenith angle tracks across the sky through the rainy season) is in fact a highly organised structure, analogous to an animal lung, exhibiting a highly organised and reticulated network. Figure 17.4(C) also shows the internal voids and channels inside a mound, which have been filled with plaster of Paris and the mound material washed away to reveal the underlying structure. These structures epitomise extremely high levels of optimisation in which a multitude of variables/constraints interact to achieve a unique solution, driven by the environmental cues found only in that particular location. Embedding and integrating this level of capability within the very fabric of our habitats may extend the process of construction and offer unique solutions in a world of climatic fluctuation and limited energy and material resources.

## 17.3 Freeform Construction Processes: A Matter of Scale

As in nature, processes and systems find niches in which they can flourish, and moving a niche system to another environment will often cause it to fail. AMT processes are no different; each process is matched to its application, which is commonly scale and resolution related. Scaling up a process, which normally deposits 10 μm beads of material to depositing 10 mm beads of material, will increase deposition rates but it will not make it possible to reproduce any of the detail possible at the smaller scale. The same applies for freeform construction, depositing 10 mm beads of material will not make it possible to define internal channels at 20 mm diameter. Here lies the crux, whatever process is developed must be able to deposit materials both at construction rates (as Pegna identified) and with the resolution of the smallest element, which must be defined within that structure. Essentially functionality may be directly related to the resolution at which fabrication takes place, no matter what scale or build rate is utilised.

Current work, both in the USA and UK, is focusing, and has been derived from, processes that would be familiar within the context of RP, i.e. extrusion-based methods and powder bed deposition systems. Though these processes may endure, the challenge of attaining rapid deposition rates, while maintaining resolution of the smallest feature, has already begun to see new solutions. Pegna's approach of bulk aggregation *in situ* presents some interesting possibilities and is one approach. What seems to be emerging is a distinction drawn between on-site and off-site solutions which reflects the way the construction industry is currently moving.

### 17.3.1 Off-Site Processes

Off-site processes fabricate pre-assembled components or modules prior to delivery to the site for assembly. Many of the emerging Digital Fabrication methods for construction extend the practice of 'off-site' construction manufacturing through highly automated and integrated subtractive and formative techniques which mimic existing cellular or flexible manufacturing systems (FMS). As this book demonstrates, the benefits of implementing AMTs through essentially 'wet' freeform construction methods, within factory environments may bring about additional 'value added' in terms of product functionality and 'end-user' requirements, which will be equally applicable to off-site construction. Methods such as FutureHome have demonstrated this route, but AMTs for construction, producing components ranging from small subassemblies to complete wall sections, will go further.

Current research at Loughborough University is exploring processes that are comparable to scaled jetting and powder bed/binder deposition techniques. Critical to this research is to identify the 'break-even' points at which a particular process becomes viable and meets the 'bottom line' requirements for 'return on investment'. Working with Z Corp, which already utilises gypsum powders, a cost analysis was performed for a comparable and existing construction product in the form of a plaster ceiling rose. These can be mass-produced, specialist or custom-made items that have a particular application in heritage building maintenance. There are a range of costs associated with these products. Table 17.1 details three ~350 mm diameter ceiling rose products and their approximate retail value. The Z Corp 810 3DP system is capable of producing components of this size in plaster and full colour. Table 17.2 details the cost analysis for reproducing a ceiling rose similar to those listed in Table 17.1.

Machine capital and maintenance costs are from Wohlers (2004). Machine utility time and depreciation were taken from Hopkinson and Dickens (2003) and were 90 % and 8 years respectively. The material cost estimates were based on data supplied by Z Corp. Finally, exact labour cost estimates were not available and the costs, shown in Table 17.2, are based on the data given for the SLS process, which is also powder based (Hopkinson and Dickens, 2003).

**Table 17.1** Approximate retail prices for plaster ceiling rose products

| Case | Company | Product | Unit cost (euros) |
|---|---|---|---|
| 1 | C & W Berry Ltd | Mass produced, UK (Artex-Rawplug). Simple design | 22 |
| 2 | Heritage Ceilings | Stock rose, hand painted, Australia. Ornate | 103 |
| 3 | Regency Town House | Stock mould, made to order, UK. Highly ornate, multi-part | 263 |

**Table 17.2** Cost analysis for plaster ceiling rose production using three-dimensional printing (3DP)

| Part name | Data | Units |
|---|---|---|
| Build data | | |
| Number per platform | 19 | — |
| Platform build time (43 mm/h) | 14 | hours |
| Production rate per hour | 1.4 | part/hour |
| Hours per year in operation (90 %) | 7884 | hours |
| Production volume total per year | 11038 | — |
| Machine costs | | |
| Machine and ancillary equipment | 135000 | euro |
| Equipment depreciation cost/year | 16875 | euro |
| Machine maintenance cost per year | 14000 | euro |
| Total machine cost per year | 30875 | euro |
| Machine cost per part | 2.80 | euro |
| Labour costs | | |
| Machine operator cost per hour | 7.07 | euro |
| Set-up time to control machine | 120 | min |
| Post-processing time per build | 360 | min |
| Labour cost per build | 56.56 | — |
| Labour costs per part | 2.98 | euro |
| Materials costs | | |
| Volume of each part | 0.003 | $m^3$ |
| Build material cost per unit volume | 36657 | euro/$m^3$ |
| Binder cost per unit volume | 49427 | euro/$m^3$ |
| Cost of material per build | 2089 | euro |
| Cost of binder per build (30 % material volume) | 939 | euro |
| Material cost per part | 159.37 | euro |
| Total cost per part | 165.15 | euro |

From Table 17.2, assuming an overhead, profit and retail mark-up of 40 %, the product cost would be 231.21 euros. The material cost is clearly significant in determining the part cost. If the plaster and binder costs could be reduced from 36657 and 49427 euro/$m^3$ to 3755 and 7510 euro/$m^3$ respectively (which is a closer approximation for the cost of stock gypsum), the resultant part retail cost would be just 34.38 euros. These values give a window in which to consider the economic case. Currently, existing 3DP products are unlikely to compete with mass production processes but could, with modifications specific to their application, compete in the near future.

### *17.3.2 On-Site Processes*

Developments for on-site or 'direct' freeform construction processes are at the most advanced stages of development at the University of Southern California under Khoshnevis (2004). Aimed at fabricating directly, on the construction site and at a range of scales simultaneously, solutions are under development for which the industry is already heralding and waiting with baited breath. Khoshnevis has been developing processes that incorporate the widespread implementation of robotic manipulators designed to handle a range of tasks, from the construction of the main structure to the placement of tiles and decorative finishes. At the heart of this comprehensive move towards construction robotics is the contour crafting (CC) process (Khoshnevis *et al.*, 2001a).

Derived from early work for an automated extrusion based method (Khoshnevis *et al.*, 2001b), contour crafting was unique in that a wide variety of materials could be extruded, including pastes, which were shaped or contoured by a secondary manipulator, or trowel, mounted on an end effector around an extrusion head. This clever solution achieved two significant advances over the fused deposition modelling (FDM) process. The process allowed for materials to be deposited that do not go through as rapid a phase change as thermoplastic extrudates. The rapid solidification of the polymer bead in FDM prevents the bead from slumping after deposition. Were a thixotropic paste to be used, for example, then slumping and spreading would occur, which is where secondary shaping comes into its own. This solution also addresses the key problem, highlighted when depositing larger amounts of material while maintaining the minimum feature resolution for which the process is required. Having a secondary shaping process, as shown in the shaped form of Figure 17.5, allows the process to be scaled while maintaining a minimum feature resolution. Typically a three-dimensional structure may be extruded on the millimetre scale while trowelling refines the structure at the micrometre scale.

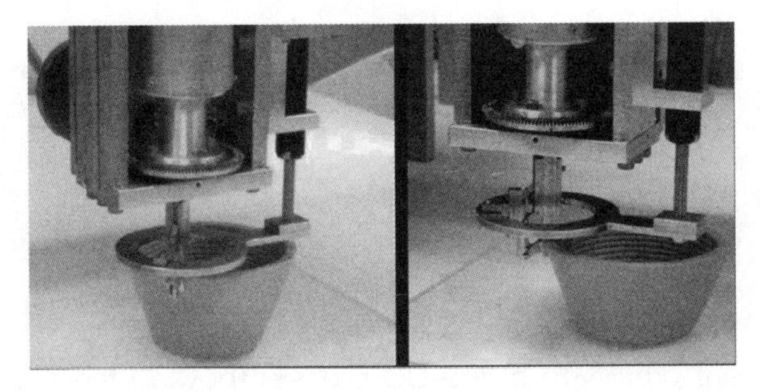

**Figure 17.5** Fabrication of contoured structures with secondary finishing. (Courtesy of Behrokh Khoshnevis, 2004)

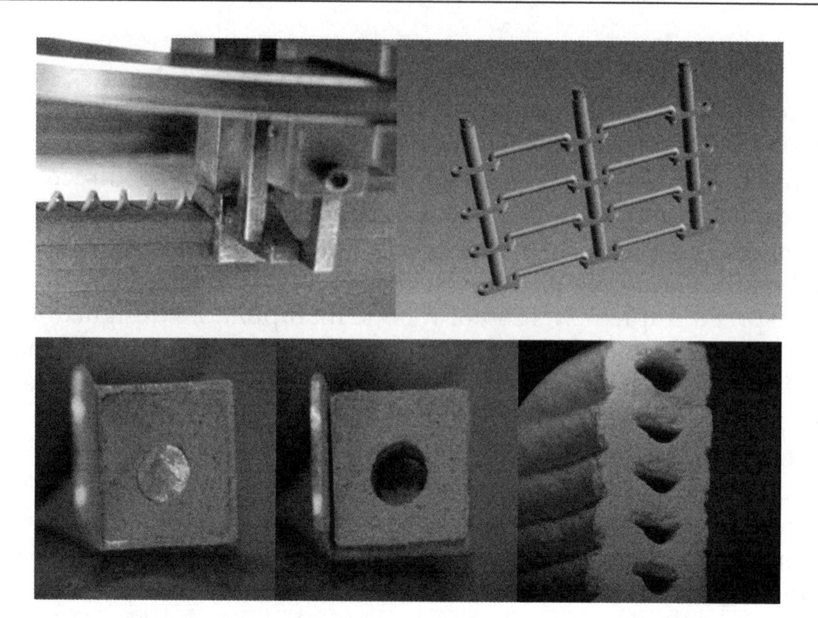

**Figure 17.6** Examples of integrated and embedded forms within contour crafted structures. (Courtesy of Behrokh Khoshnevis, 2004)

Contour crafting is an ideal candidate for modifying the extruded material in line with the functional requirements of the structure. The process has demonstrated it ability to not only deposit reinforcing media extruded simultaneously with the build matrix but also to allow the inclusion of shaped mandrills to control permeability, as shown in Figure 17.6. Work is currently under way to incorporate secondary embedded elements, reinforcing subassemblies, integrated services and distribution networks within a structure during deposition.

The research is taking place on many fronts simultaneously. Current work is exploring the problem of the speed of deposition combined with high surface definition in a similar way that many existing AMT processes exploit a 'skin' and 'core' strategy, as shown in Figure 17.7. With CC the approach is

**Figure 17.7** Current prototype of the CC process showing the high-definition 'skin' and infill 'core' strategy. (Courtesy of Behrokh Khoshnevis; the right-hand image is from Khoshnevis, 2004)

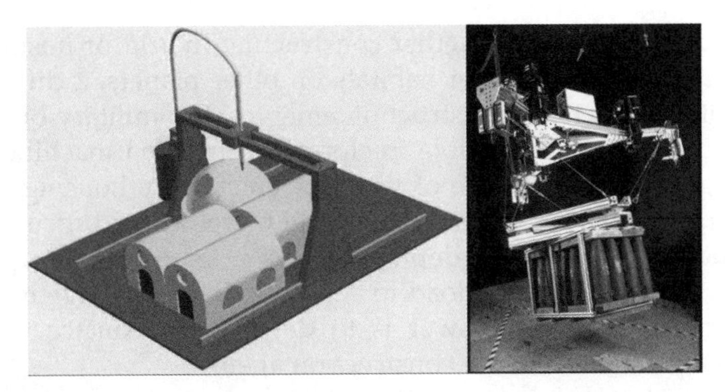

**Figure 17.8** Gantry type and RoboCrane delivery systems for the CC process. (Courtesy of Behrokh Khoshnevis, 2004)

to utilise high-definition extrudates for the external faces of the structure, while using a generic 'infill' strategy for the internal elements, analogous to applying contoured shuttering for a concrete infill. It goes without saying that materials development and characterisation are critical to the success of the process and a larger variety of material solutions are being investigated in conjunction with Degussa GmbH.

Looking beyond material delivery and end-effector design, Khoshnevis (2004) is considering placement devices such as those shown in Figure 17.8. Work has been done with both standard gantry systems, non-Cartesian robotics and recently, with the US National Institute of Standards and Technology (NIST), RoboCrane.

RoboCrane is a much needed development in terms of large-scale controllable robotics. By utilising suspended platforms and retractable wiring, the system is both scalable and deployable and should allow for multiple platforms to work in close proximity. The project will use a platform with a CC nozzle installed to look at what level of accuracy can be considered. The project also looks beyond terrestrial applications and will suit reduced gravity applications.

### *17.3.3 Off-World Processes*

One final attribute which will come from freeform construction solutions is that of deployment. As we have seen, skyscraper manufacturing systems demonstrate that truly enormous gantry robotic systems are possible, even for the largest structures on earth, but there are, however, applications where such large machines cannot be considered.

Assuming that freeform construction may allow habitats to be built that can be de-coupled from the requirements to be near energy supplies or waste management systems, through the integration of self-regulating control systems within the structure, then deployment becomes one of the overriding considerations. Whether constructing in arid or hostile environments on earth or constructing habitats on other planets, a different set of requirements for freeform construction emerges. For military bunker applications, a single large deployable freeform construction machine would be an easy target within a theatre of war. For emergency housing solutions a single freeform construction machine would therefore be difficult to load on to a transporter or plane for deployment and, for off-world applications, there is physically not the payload to accommodate a massive machine. For all these applications the answer is to devolve the construction process between many smaller mobile construction agents.

As to the level of devolvement, the solutions emerging draw out interesting possibilities and may offer a glimpse as to the direction AMTs will take in the future. Figure 17.9 considers the transport and assembly of robotic assembly equipment in which construction occurs at many scales simultaneously. In this approach construction occurs on the macro-scale with a gantry-type CC machine (Khoshnevis and Bekey, 2002) fed by a material umbilical from a lunar regolith processing plant. *In situ* resource utilisation (ISRU) will be paramount to any off-world or, for that matter, remote terrestrial applications. Servicing this machine, as well as performing material excavation and pre-processing, are smaller robotic devices. Certainly, for a lunar application, the control and coordination could be performed remotely through telepresence systems based around operators on earth.

For situations outside the line of view of a radio link, and particularly where a signal may take 8 minutes to travel from Earth and 8 minutes back,

**Figure 17.9** Deployable CC lunar application. (Courtesy of Khoshnevis and Bekey, 2002)

such as Mars for example, then construction agents are required to be capable of coordinating and making their own decisions. In addition, the types of habitats required for a Mars colony will need the levels of adaptability possible through structural homeostasis, and even this will not be enough. These structures will need to be self-sufficient and be able to optimise themselves dynamically for extreme climate fluctuations, for ground movement and for increasing numbers of colonists as they arrive on Mars. For this we will need agents that can:

1. Construct a habitat.
2. Inhabit the structure.
3. Maintain all environmental conditions within stable equilibria.
4. Optimise the structure specifically for the inhabitants through generation and degeneration of the structure and form.
5. Recycle and manage waste streams.
6. Propagate and supply nutrients to the inhabitants.
7. Regulate respiratory gas requirements between humans and plants.

Such a system so resembles an organic system that it will be difficult to distinguish the two and the solution also lies with the social insects. Researchers at Loughborough University and the University of Nottingham are unravelling how such complex levels of organisation, leading to dynamic three-dimensional structures, are performed by social insects. With almost no discernable brain and no master plan, how do they coordinate their activities to achieve complex tasks?

At its core lies a process known as stigmergy. Stigmergy has many definitions, but essentially complex structures emerge through the seemingly uncoordinated work of many individual insects driven by environmental cues triggered by pheromone dispersal. Pheromones are scents, very similar to solvents, laid down by social insects during activity. Being a solvent, pheromone evaporates at an exact rate, matched precisely to a particular species of social insect and, more importantly, matched to the speed at which they work. In the case of the termites considered for structural homeostasis, whatever activity is performed is always followed by the deposition of a pheromone, with different pheromones being used for different tasks. Termites are tuned to their environment and seek to modify their surrounding physical conditions to what they interpret as 'correct'. In the case of construction, worker termites seek out a set of environmental conditions that make them 'feel' comfortable, which include moisture levels, $CO_2$ and $O_2$ concentrations and, in some species, temperature. If the optimum changes through, say, damage, then they are driven into action simply as a response to a change in the environment.

Figure 17.10(B) shows this process underway. Worker termites continuously mix and grade construction materials (from minerals mined deep

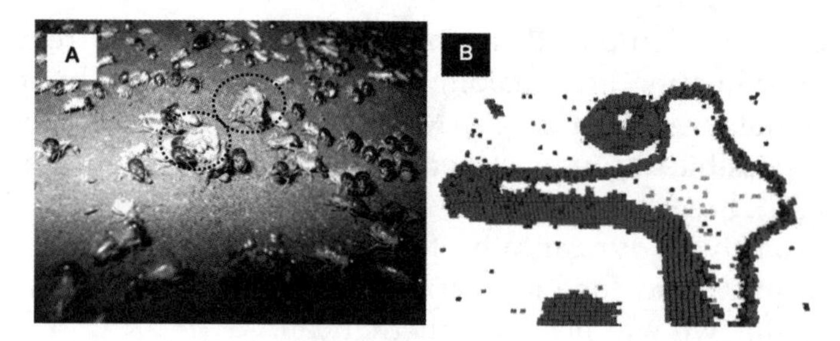


**Figure 17.10** (A) Initiation of mound construction driven by stigmergic cues (pillars are circled). (Courtesy of Rupert Soar. (B) Stigmergic artificial intelligence (AI) simulation showing construction agents utilising termite-derived algorithms for the three-dimensional structure. (Courtesy of Rupert Soar and of Bai Li at hottingham university)

underground and bound with polysaccharide derivatives) but do not deposit packages of build material at random. The first termite on the scene will sense and move towards a point, where a balance of the environmental cues, normally $CO_2$ derived in the form of a gas gradient, is located. If it finds this combination it will place its package on the ground and, just before depositing it, will mix in pheromone and leave. If the environmental optimum persists then the next termite comes along and, instead of making its own decision where to place the build material on the ground, its decision will be overridden by the presence of the pheromone from the first package deposited. This method of positive feedback or reinforcement results in the construction of pillars, which at standing height are formed into arches which become a corridor. This is known as the 'primary build' and does not result in a homeostatic structure. The primary structure must then be modified to form the large chambers and structure associated with the mound, driven by secondary stigmergic cues such as ventilation, communication and transportation within the macro-structure.

What can we learn about freeform construction applications? One of the largest problems to be faced with the implementation of collaborative or swarm construction agents will be the level and mechanism of coordination that results in adaptive homeostatic three-dimensional forms. Assuming that mass producible construction robotics are only a decade away, the challenge will be how to coordinate them on a massive scale. Attempts to direct each and every agent (by embedding a complete internal environment representation), and the order in which they deposit build material, will encounter the 'bandwidth' problem. This is a known limit to how much data can be transmitted to all agents simultaneously and, for a structure as complex as a homeostatic habitat, no level of computing power will allow for this level of control in the classic hierarchical sense of the term. The construction agents shown in Figure 17.10(B) execute a series of just five hierarchical decisions:

1. Block interaction in response to block collision at time $t - 1$.
2. Block interaction in response to queen pheromone $Q$.
3. Block interaction in response to $C_t$.
4. Movement in response to cement pheromone $C_{t-1}$, $C_t$, $C_R$ and $C_L$ (where R and L stand for right and left directions respectively).
5. Random walk.

All the information relating to the environment is drawn directly from the environment itself and therefore none of this detail is required in the program. Some of the environmental information is derived from the presence of the pheromones, which dictate the final form, and some are drawn from the physical environment. Imagine how a prevailing wind will interact with the dispersion of the build pheromone. As it blows through the construction site, it will change the shape of the pheromone dispersion gradient and the agents will immediately build an optimised new gradient, resulting in an optimised structure without ever programming in any new information into the agent. Importantly, however, the agents seek conditions that fall within upper and lower threshold limits. Once an enclosed structure is formed, the concentration of pheromones will rise, which induces the agents to remove build material to expand the structure so that the pheromone concentrations are correct for an enclosed structure.

The ability to optimise a component intrinsically, for its specific use or environment, has important implications for AMTs. In the same way that 'bandwidth' is known to be a limitation in AI/robotic systems, the same is also true for optimised functionally graded components. Moving to voxel-based approaches introduces similar bandwidth problems when calculating the specific location and combination of materials. At the micrometre and ultimately the nanometre scale the production of truly dynamic structures, on a par with osteocyte optimised trabecular bone architectures, will emerge. Unfortunately, by then, we will need to rethink our definition for additive systems as they will probably have to include subtraction within the same process.

## 17.4 Conclusions

AMTs for construction are a new and exciting extension of the technology into a new field. Though construction is, inherently, an additive process, it is automation, design freedom, integrated services and client-led design that the construction industry is keen to explore. Current state-of-the-art in construction currently focuses on modular solutions, either fabricated directly on-site or as pre-assembly solutions off-site. AMTs offer a 'third way' through the implementation of 'direct' fabrication methods by incorporating material preparation, delivery and deposition into a single operation.

Construction is a highly cost/time-driven industry, for which AMTs will have to compete. However, there are key opportunities for exploiting 'added value', through increased functionality above and beyond existing construction capabilities. In particular, the opportunity exists to make the link between computational-based structural optimisation and the physical processes that will be required to reproduce these structures, for which the term freeform construction is being applied.

Research goes back to 1997 with the first methods for which AMTs could be applied for construction were considered. This work began to highlight the specific requirements of the construction industry, were AMTs to be applied, and produced solutions for a selective aggregation process.

Active research is now under-way in the USA and UK to explore the process requirements for the many niches for which freeform construction processes will be applied. The research is diverse and covers processes such as contour crafting, which is rapidly approaching commercialisation, as well as technologies specific for construction materials specialists, such as Degussa GmbH and BPB plc, who realise the importance of the technology. Researchers at Loughborough, Nottingham and Cambridge Universities in the UK and USC and Syracuse University in the USA are realising the full potential of the technologies in the light of drivers such as climatic variation, energy shortages, skills shortages and waste management issues, which will hit the construction industry harder than most. Novel solutions for 'single material' structures, fully integrated services, self-supporting geometries, 'stealth' structures, highly complex homoeostatic habitats and autonomous construction agents promise to offer the 'step changes' to the construction process for which many are currently seeking answers.

## References

1. Burns, M. (1993) *Automated Fabrication, Improving Productivity in Manufacturing*, PTR Prentice-Hall, Englewood Cliffs, New Jersey.
2. Gibb, A.G.F. (2001) *Pre-assembly in Construction: A Review of Recent and Current Industry and Research Initiatives on Pre-assembly in Construction*, CRISP Consultancy Commission, 00/19.
3. Guthrie, P., Coventry, S., Woolveridge, C., Hillier, S. and Collins, R. (1999) *The Reclaimed and Recycled Construction Materials Handbook*, CIRIA, London.
4. Hopkinson, N. and Dickens, P. (2003) Analysis of rapid manufacturing – using layer manufacturing processes for production, *Proceedings of the*

*Institution of Mechanical Engineers, Part C: Journal of Mechanical Engineering Science,* **217**(C1), 31–40, Professional Engineering Publishing Ltd.

5. Howe, A.S. (2000) Designing for automated construction, *Automation in Construction,* **9**, 259–76, Elsevier Science.
6. Khoshnevis, B. (2004) Automated construction by contour crafting – related robotics and information technologies, *Journal of Automation in Construction – Special Issue: The Best of ISARC 2002,* **13**(1), January 2004, 5–19.
7. Khoshnevis, B. and Bekey, G. (2002) Automated construction using contour crafting – applications on earth and beyond, in Proceedings of the 19th International Symposium on *Automation and Robotics in Construction,* Gaithersburg, Maryland, pp. 489–94.
8. Khoshnevis, B., Bukkapatnam, S., Kwon, H. and Saito, J. (2001a) Experimental investigation of contour crafting using ceramics materials, *Rapid Prototyping Journal,* **7**(1), 32–41.
9. Khoshnevis, B., Russell, R., Kwon, H. and Bukkapatnam, S. (2001b) Contour crafting – a layered fabrication technique, *Special Issue of IEEE Robotics and Automation Magazine,* **8**(3), 33–42.
10. Pegna, J. (1997) Exploratory investigation of solid freeform construction, *Automation in Construction,* **5**(5), 427–37.
11. Tsui, E. (2002) *Evolutionary Architecture: Nature as a Basis for Design,* John Wiley & Sons, Ltd, Chichester. ISBN 0471117269.
12. Wing, R. and Atkin, B. (2002) FutureHome – a prototype for factory housing, in 19th International Symposium on *Automation and Robotics in Construction (ISARC),* National Institute of Standards and Technology, Gaithersburg, Maryland, 23–25 September 2002.
13. Wohlers, T. (2004) *Rapid Prototyping, Tooling and Manufacturing: State of the Industry,* Wohlers Associates, Colorado.

# 18

# Rapid Manufacture for the Retail Industry

Janne Kytannen
*Freedom of Creation*

## 18.1 Introduction

This book has been largely written by practitioners in the field, most of whom are engineers. In this chapter the experience and perspective of creative designers is put forward in order to provide a wider view of the current uses and longer-term potential of Rapid Manufacturing (RM). The chapter covers a range of topics from lighting products that are already being sold through retail outlets through to the vast possibilities that may be achieved by RM.

## 18.2 Fascinating Technology with Little Consumer Knowledge

For most people who have seen Rapid Prototyping (RP) machines, the initial introduction to the technology has resulted in a myriad of future possibilities flashing through their minds. Although we are still a long way from building parts atom by atom, the near future's nano-assemblers might be able to create objects additively from a variety of different materials simultaneously.

Technologies such as stereolithography (SL) have already been available for over 15 years, yet still only a small percentage of people are aware of them. It is remarkable that such intriguing technologies have not yet found

*Rapid Manufacturing: An Industrial Revolution for the Digital Age*
Editors N. Hopkinson, R.J.M. Hague and P.M. Dickens 

their way into consumer knowledge. Why has an attempt not been made to bring the technologies to the forefront of consumer knowledge? Perhaps there have been many attempts to achieve this, but have none ever been developed seriously?

## 18.3 The Need for Rapid Prototyping to Change to Rapid Manufacturing

It is broadly accepted that in order for the RP/RM industry to flourish it needs to attract consumers by using these machines for the manufacturing of end-use products for the masses. Has there been sufficient effort applied to achieve this? It is likely that most of the industry's leaders would respond that neither the machines nor the materials are ready for that yet. The view taken at Freedom of Creation (FOC) is that this is an incorrect opinion simply for the reason that every technique and every material has its own characteristics. New ideas have always been alien and require time and the right application to be embraced. Have creative minds with a totally different mind-set to the technically minded majority of RP/RM users ever been hired to ponder the true value and potential for these technologies? Current materials and machines are simply created to mimic existing products on the market and thus serve an already existing market. From the perspective of a designer, this is very far from pushing the envelope, something the industry truly needs to do in order to flourish beyond RP.

It is difficult to picture these machines producing products that have not even existed before. It is even more difficult for the industry to invest money into something that nobody is yet able to comprehend. However, this has to be done if the industry wishes to one day see these machines in people's homes, where they could download the three-dimensional files of the products of their choice. Current machines can already produce the wildest ideas, but we simply have not been able to experiment widely enough and thus come up with more clever applications for their usage yet.

## 18.4 Rapid Manufacturing Retail Applications

### *18.4.1 Lighting*

When FOC first designed RM lighting concepts at the end of the 1990s there were still several technical hurdles to overcome, as well as the need to convince people to invest into the necessary R&D and commercialisation. Consequently, it took a few years before the lighting concepts became commercial products.

Initially it sounded ridiculous to create lights from SL resins, which are intended to maintain their form and colour for only a few months. One of the

**Figure 18.1** Lampshade made by stereolithography. (Reproduced by permission of Freedom of Creation Group)

first ideas to deal with this was to see what happened if the curing process was accelerated using excessive heat or ultraviolet (UV) light. After experimenting with different resins, it was possible to stabilise beautiful colours for RM lighting objects, which looked more like amber than plastic (see Figure 18.1).

Another hurdle was to overcome the economics of the RM processes. It was important to find ways to make objects in a manner that would maximise the usage of the build platform, minimise the materials needed, reduce labour costs to a minimum, yet still make commercial design products. Several ideas surfaced on how to make products with self-supporting structures for SL or collapsing and nesting products for selective laser sintering (SLS). Figure 18.2 shows a series of nested lampshades made by SLS.

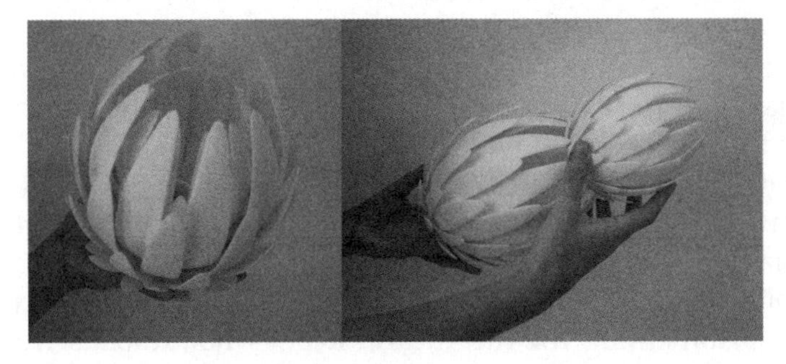

**Figure 18.2** Nested 'Lily' and 'Lotus' lampshades made by SLS (Reproduced by permission of Freedom of Creation Group)

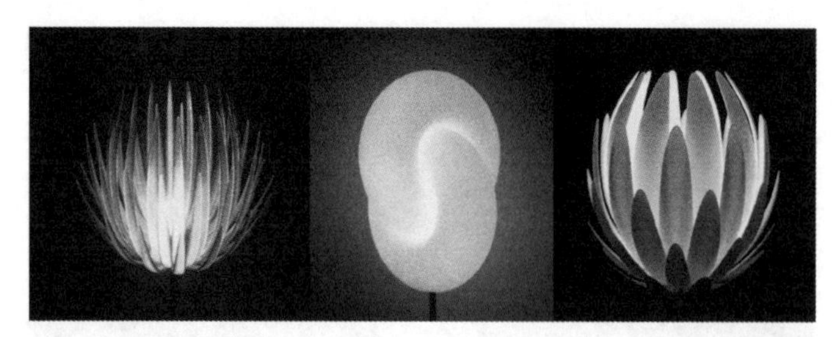

**Figure 18.3** Lotus and open cube lampshades (Reproduced by permission of Freedom of Creation Group)

Materialise was the first company to invest into the research and development of this first series of lighting products. The products retail between 300 and 1200 euros and global commercialisation started in October 2003. Figure 18.3 shows examples of different lighting products that are currently being sold from 30 retail outlets worldwide.

In a very short period of time, the designs and methods used to produce these products have gained huge public interest and they have already been published in numerous designs publications. *Icon Magazine* was the first to publish an editorial article on FOC, June 2003, titled 'Lampshades that will change the world'. In January 2004 Elle Decoration UK selected two of the designs at their annual design review as the Number 1 design buys for 2004.

### *18.4.2 Three-Dimensional Textiles*

#### The Difficult Acceptance of Something Alien

The most controversial ideas are always the most fun to develop, but their introduction often takes quite some time. FOC first presented the possibility of three-dimensional printed fabrics to a few RP companies in 1999; however, it took a couple of years from the initial concept before anyone even gave it a try.

When one talks about textiles people always think of cotton, wool or other soft materials that they can relate to. Then they make the comparison with the current RP materials and make the assumption that the same could never be done with these techniques. Regrettably this is a typical but somewhat limited view and this type of thinking will hardly lead into anything new or controversially different.

Even though making textiles with similar detail as with today's weaving techniques is something for the future, it does not stop us from thinking about other products we might be able to make using today's technologies.

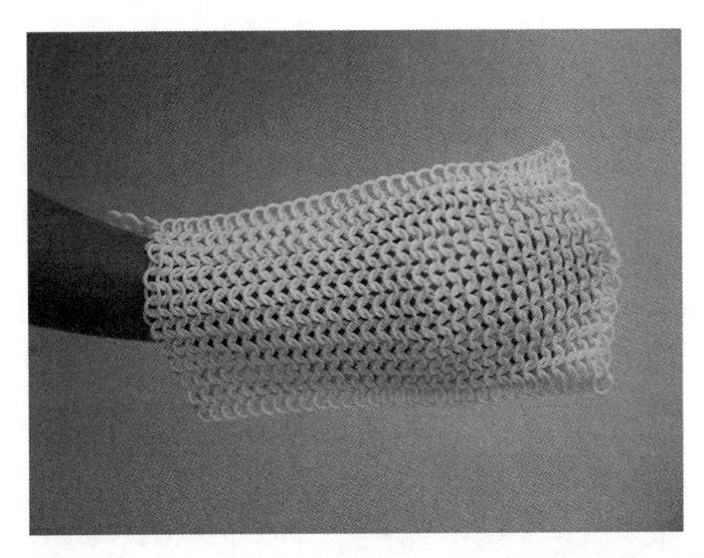

**Figure 18.4** Shower scrub made by SLS (Reproduced by permission of Freedom of Creation Group)

The first step is purely to see what kinds of tools we have at hand, what kind of materials we have and simply come up with something feasible with them.

FOC received a lot of criticism from people about the roughness of SLS polyamide, which was first used for making patterns for fashion accessories. Apparently it was not nice to the skin and the resolution was not fine enough. Inspired by this, the process was used to make functional shower scrubs instead (see Figure 18.4).

## Commercial Applications

When comparing lighting products with textiles there is a fundamental difference in terms of manufacturing logistics. With lighting products a complete assembly including a bulb, electrical cables, base, etc., is required along with CE approval. All of these issues incur some labour and storage costs. In fact only one part of the product is produced by RM. Products that can be manufactured out of one material in one assembly are the ideal first step for great RM products. FOC's first line of handbags for the fashion industry require no storage, nor assembly, but already consist of hundreds of separate parts from rings to zippers, to hinges to straps and functional clicking systems. The only post-process these products require is the right tumbling process, which gives them the required smooth surface finish. Figure 18.5 shows a section of 'fabric' made by SLS used for Rapid Manufactured handbags.

**Figure 18.5** Section of 'fabric' made by DMLS used for handbags (Reproduced by permission of EOS GmbH)

## 18.5 Mass Customisation

### *18.5.1 Mass Customised Retail Products*

Mass customisation has been gien a lot of attention over the last decade. Some companies have already been able to successfully introduce retail products to the market that enable the consumer to change some aspects of their products. Nike, for example, is selling shoes through their website, where consumers have the opportunity to alter the colours, size and add text such as initials to the product. Similarly Dell has boomed with tailor-made computers sold through their website.

### *18.5.2 Future Possibilities of Mass Customised RM Products*

Even though RP/RM machines are on a totally different level when it comes to enabling personal freedom of choice for tailor-made products, we have much more than technology to consider. What features would people be able to alter before the product would lose the designers signature? Changing the size, colour or material properties is quite easy, but would the designers want consumers to create totally bespoke products for their own liking? However, it is likely that in the future products that can most easily be changed will be the ones that will thrive.

Mass customisation raises a lot of questions, which all are vital to explore. Currently, design freezes are required in order to ensure that products reach the consumer. However, if technology allowed us to create any object at any time at the touch of a button, would design freezes become a thing of the past? Would people be tempted to re-design their possessions on a daily basis to achieve continued improvement? Would this affect the way in which people perceive and value their material possessions? As designers, FOC have faced a completely new hurdle to overcome when using RP/RM tools for manufacturing purposes. In other design work constraints have always been applied by the technologies for manufacture. Of course, there are constraints associated with RP/RM, but the gap is rapidly narrowing and perhaps we will one day be in a situation where we can make anything out of anything. This actually presents a major question: 'What to do if you can make anything you wished for?'

Mass customisation gives us some clues of what the future might hold for product development within RM. The accelerating pace of designing, prototyping and manufacturing give us more variety of products each year, when compared to the previous. The days of huge factories producing millions of similar products are bound to perish if others are able to adjust to the people's personal wishes and changing trends with new manufacturing logistics. For example, a company X is developing and commercialising sunglasses by using old-fashioned manufacturing logistics and injection moulding tools. An average company would be able to produce two different lines of sunglasses a year, in order for it to be economically feasible. Here is an example of an alternative process using RM means. FOC were assigned by Heineken to develop a business gift for the 10 year anniversary of their cooperation with Dance Valley (the biggest annual outdoor dance event in The Netherlands). Only one item was produced, no prototypes were required and the development time from concept to a final. STL (Standard Triangular Language) file took only 10 hours. According to FKM Sintertechniek, who produced the product, the sintering time for this item was 8 hours. The following is hypothetical, but let us imagine that you would have annually 200 hundred similar projects. By using this approach you would only need to produce a few hundred and sometimes only dozens of each product in order to get to the same kind of production quantities as our mass-production counterparts. However, since the products can be unique, you are creating a totally new service that never existed before and an unlimited range of products for your catalogue.

### 18.5.3 Limitations and Possibilities

Clearly it will take a long time to gain the vast expertise with the machines, so that a computer model on the computer screen will perform exactly the way that was intended in real life without any requirements for prototyping. However, if we can do this with a great range of lighting designs already, there will be a lot of other designers in the future who will be able to do it as well and hopefully find even better applications for their usage.

## 18.6 Experimentation and Future Applications

Experimenting is the key for any future development in Rapid Manufacturing. Pretty much all of FOC's designs for RM have started from a hunch and pure curiosity for experimenting something new. Many people ask about the possible future applications for RM. Since we are talking about a technology, which is only starting to find its way to manufacturing consumer goods, the direction can be pretty much anything. For example, MIT claims that their invention of three-dimensional printing will be able to print any material, so the more suitable question might be: 'Where would these tools not give us an advantage in the future of manufacturing?'

# Index



*Rapid Manufacturing: An Industrial Revolution for the Digital Age*
Editors N. Hopkinson, R.J.M. Hague and P.M. Dickens.